T0073189

Introduction to
Algebraic Coding Theory

Contemporary Mathematics and Its Applications: Monographs, Expositions and Lecture Notes

Print ISSN: 2591-7668
Online ISSN: 2591-7676

This series aims to inspire new curriculum and integrate current research into texts. Its aims and main scope are to publish:

- Cutting-edge Research Monographs
- Mathematical Plums
- Innovative Textbooks for capstone (special topics) undergraduate and graduate level courses
- Surveys on recent emergence of new topics in pure and applied mathematics
- Advanced undergraduate and graduate level textbooks that may initiate new directions and new courses within mathematics and applied mathematics curriculum
- Books emerging from important conferences and special occasions
- Lecture Notes on advanced topics

Monographs and textbooks on topics of interdisciplinary or cross-disciplinary interest are particularly suitable for the series.

Published

Contemporary Mathematics and Its Applications
Monographs, Expositions and Lecture Notes

Vol. **3**

Introduction to Algebraic Coding Theory

Tzuong-Tsieng Moh
Purdue University, USA

World Scientific

NEW JERSEY · LONDON · SINGAPORE · BEIJING · SHANGHAI · HONG KONG · TAIPEI · CHENNAI · TOKYO

Published by

World Scientific Publishing Co. Pte. Ltd.

5 Toh Tuck Link, Singapore 596224

USA office: 27 Warren Street, Suite 401-402, Hackensack, NJ 07601

UK office: 57 Shelton Street, Covent Garden, London WC2H 9HE

Library of Congress Cataloging-in-Publication Data

Names: Moh, T. T., author.

Title: Introduction to algebraic coding theory / Tzuong-Tsieng Moh, Purdue University, USA.

Description: New Jersey : World Scientific, [2022] | Series: Contemporary mathematics and
its applications: monographs, expositions and lecture notes, 2591-7668 ; Vol. 3 |
Includes bibliographical references and index.

Identifiers: LCCN 2022000598 | ISBN 9789811220968 (hardcover) |
ISBN 9789811220975 (ebook for institutions) | ISBN 9789811220982 (ebook for individuals)

Subjects: LCSH: Coding theory.

Classification: LCC QA268 .M64 2022 | DDC 003/.54--dc23/eng20220225

LC record available at https://lccn.loc.gov/2022000598

British Library Cataloguing-in-Publication Data

A catalogue record for this book is available from the British Library.

For any available supplementary material, please visit
https://www.worldscientific.com/worldscibooks/10.1142/11849#t=suppl

Desk Editors: Soundararajan Raghuraman/Lai Fun Kwong

Typeset by Stallion Press
Email: enquiries@stallionpress.com

Printed in Singapore

TO

My wife Ping

Preface

This is a compilation of lecture notes from a course for second-year graduate students at Purdue University, who are only required to have some background in abstract algebra. The development of algebraic coding theory correlates to the development of algebra and with a delay of about one hundred years. Together, the lecture notes form the following four parts in this book: Part I is about the elementary theory of algebraic coding. Mathematically, the tools used are vector spaces and linear algebra. Part II is about various rings and the associated algebraic codings. A fast decoding method is presented. Part III is a survey of the useful parts of algebraic geometry. Part IV is about geometric coding theory.

The ultimate goal of this text is to introduce the students to the modern *geometric Goppa codes*. Since the elementary coding theory is assumed to be of interest only to students in engineering or computer science and hence unknown to many students in mathematics, we will begin with some discussions of this elementary material. It is naturally progressed to the ring-theoretic codes, mainly BCH codes, Reed–Solomon codes, and classical Goppa codes, which can easily be discussed in the context of polynomial and rational function rings and formal power-series rings. A complete treatment of decoding ring codes is given in Part II.

One of the challenges is to present a survey of the useful parts of algebraic geometry, which is an elegant and important topic in mathematics. There are many ways to teach the subject. In the chapter on algebraic geometry, Part III, we adopt Chevalley's algebraic approach to the curve theory and provide many examples to illustrate the theorems. Many of the proofs are consigned to classic books by Zariski and Samuel [16], Walker [15], Chevalley [10], and Mumford [13].

The final part of this text is devoted to the well-known geometric Goppa codes [25]. Their early decoding processes depend on linear algebra only and are elementary. For the decoding processes, we give explicit descriptions and try to present them as naturally as possible. Further discussions involving the remarkable concepts of Feng and Rao [21] on majority voting will also be presented in this way.

The field of coding theory is too rich to be covered in a one-semester course. We have added appendices to discuss the familiar topics such as *convolution codes, sphere-packing problem,* other interesting *codes,* and *Berlekamp's algorithm* which might be beneficial to interested readers who wish to have a wide scope of understanding of the related materials.

This book is written to be concise. There are about 204 pages for a one-semester course. We hope that it will be useful to students and working algebraic geometers alike in understanding the booming field of coding. We would like to thank W. Heinzer for reading the whole book and making some valuable suggestions and B. Lucier for commenting on Parts I & II of our manuscript.

We wish to thank Ms. Rochelle Kronzek, executive editor at World Scientific Publishing Company, for her constant enthusiasm on initializing this project. We are grateful to Ms. Lai Fun Kwong, managing editor at WSP, for her prompt communications and support during the book writing. We wish to thank the anonymous referee who improved this book and Mr. T. R. Soundararajan for taking care of the final form of the book.

About the Author

T. T. Moh is an Emeritus Professor of Mathematics at Purdue University, specializing in algebra and algebraic geometry. He received his PhD in mathematics from Purdue University in 1969 and became an Assistant Professor there afterward. He also spent time at the Institute for Advanced Study in Princeton and was on the faculty of University of Minnesota before rejoining Purdue University in 1976.

Contents

Appendices 207

Introduction

All living beings use signals to communicate with each other. The signals, also known as codes, can take the form of chemicals, sounds, colors, etc. About two million years ago, humanity gained its own distinctiveness by creating abstract signals, languages. All languages can be seen as codes. Many historians try to decipher "lost" languages: the most famous one was probably written hieroglyphs, which were deciphered using the Rosetta stone. Since ancient times, poems have been used as a way of communicating oral tradition, including *Illiad* and *Odyssey* in Greek, *Mahabharata* and *Ramayana* in Sanskrit, and *Ode* in Chinese. One advantage of the rhyme and poetic structure of poetic verse over prose is that it is easy to find the errors if they occur. In other words, poetry is the first "error-detecting" form of communication.

In 1945, Erwin Schrödinger published a book entitled *What is life? The Physical Aspect of the Living Cell* [3], in which he observed that chromosomes are code-scripts and are molecules in nature. Ergo, there must be a code of some kind which allowed molecules in a cell to carry information. It motivated many scientists to study the codes transmitted by living beings, eventually leading to the discovery of the double-helix structures of DNA and RNA by James D. Watson and Francis Crick.

Poetry uses rhymes and molecules use chemical bonds to detect errors and correct them. It is natural to impose some algebraic relations on the symbols of letters for the same purpose. A computer scientist, R. W. Hamming, used a primitive computer, by today's standard, to perform his research. At that time, scientists had to queue their work for the computer to sequentially process. If the computer found errors (usually the errors were caused by misreading the program by the computer) in the program, it would skip the task and proceed to the next one in the queue. The researcher would have to correct the errors in the faulty program if there is a mistake, in any case resubmit it to the queue, and wait several weeks for the computer to find time to work on it again. Hamming was annoyed by the wait and decided to create a code to prevent the computer from misreading the program. His invention of "self-correcting code" introduces some vector-space structures for words. This is a great invention and was named after him. Let us introduce his idea as follows.

Let us consider $GF(2)$ $(=\mathbf{F}_2)$, the field of two elements $\{0,1\}$ which will be called the set of *letters*, and message $a_1a_2a_3a_4$, where $a_i \in \{0,1\}$. Hamming added three more symbols $b_1b_2b_3$ by the following formula:

$$b_1 = a_1 + a_3 + a_4,$$
$$b_2 = a_1 + a_2 + a_3, \tag{1}$$
$$b_3 = a_2 + a_3 + a_4.$$

Then, Hamming used seven symbols $a_1a_2a_3a_4b_1b_2b_3$ to carry the message of four symbols $a_1a_2a_3a_4$. We may consider the following matrix multiplication, with

$$G = \begin{pmatrix} 1 & 0 & 0 & 0 & 1 & 1 & 0 \\ 0 & 1 & 0 & 0 & 0 & 1 & 1 \\ 0 & 0 & 1 & 0 & 1 & 1 & 1 \\ 0 & 0 & 0 & 1 & 1 & 0 & 1 \end{pmatrix}$$

and $[a_1a_2a_3a_4] \times G = [a_1a_2a_3a_4b_1b_2b_3]$. The matrix G is called the *generator matrix*. The a_i's are called the *message symbols*. Furthermore, let

$$H = \begin{pmatrix} 1 & 1 & 0 \\ 0 & 1 & 1 \\ 1 & 1 & 1 \\ 1 & 0 & 1 \\ 1 & 0 & 0 \\ 0 & 1 & 0 \\ 0 & 0 & 1 \end{pmatrix}$$

and $[a_1a_2a_3a_4] \times G \times H = [000]$. The matrix H is called the *check matrix* and b_i's are called the *check symbols*.

The decoding process is as follows. Suppose that the computer, for whatever reasons, reads $[a_1a_2a_3a_4b_1b_2b_3]$ as $[a_1'a_2'a_3'a_4'b_1'b_2'b_3']$ which might be different from the original string. However, this kind of error is infrequent, so we may reasonably assume that there is at most one error, i.e., either

$$[a_1'a_2'a_3'a_4'b_1'b_2'b_3'] = [a_1a_2a_3a_4b_1b_2b_3]$$

or

$$[a_1'a_2'a_3'a_4'b_1'b_2'b_3'] = [a_1a_2a_3a_4b_1b_2b_3] + [0\cdots010\cdots0].$$

The computer calculates

$$[a_1'a_2'a_3'a_4'b_1'b_2'b_3'] \times H = [c_1c_2c_3].$$

If $[c_1c_2c_3] = [000]$, then the above defining equations (1) for b_1, b_2, b_3 show that there is either no error or there is more than one error. Since we assumed that there is at most one error, we may conclude there is no error, so the computer should take the message $[a_1'a_2'a_3'a_4']$. If $[c_1c_2c_3] \neq [000]$, then we have

$$
\begin{aligned}
[a_1'a_2'a_3'a_4'b_1'b_2'b_3'] \times H &= ([a_1a_2a_3a_4b_1b_2b_3] + [0\cdots010\cdots0]) \times H \\
&= [0\cdots010\cdots0] \times H \\
&= [c_1c_2c_3].
\end{aligned}
$$

Therefore, $[c_1c_2c_3]$ must be one of the row vectors of the matrix H, and thus the computer locates the position of the error. The computer simply flips the bit at that position. In this way, the computer will correct the code $[a_1'a_2'a_3'a_4'b_1'b_2'b_3']$ and take the first four bits of the corrected message as the message. This code not only detects the error but also corrects the error. Note that the *location* of the error must be detected before it can be corrected, a process that will be referred to as finding *error locator*.

However, there might be two or more errors, in which case Hamming code fails and the above method decodes the message to a wrong word. One may assume that the possibility is rather small. The usage of Hamming codes is to eradicate a single error, but it is ineffective if the errors are multiple.

For a longer message, we can chop it into blocks (each with four bits), padding the end if necessary. This produces a *block* code.

The principles of self-correcting in Hamming codes are valid today, and widely used in communications through noisy channels. Academically, **code** means **self-correcting code**. All channels of communications are noisy to different degrees. The self-correcting codes have become prominent today. Furthermore, although the above Hamming code is merely a simple exercise in linear algebra, the implications were groundbreaking.

As important as Hamming's work, around that time, C. E. Shannon was developing his theory of information. Shannon wanted to know how we measure *information*. Shannon took the idea of *entropy* from thermodynamics, which is defined to be *logarithm of disorder*. The second law of thermodynamics states: *The entropy of an isolated system is always increasing*, for instance, if one mixes cold water with hot water, then one gets warm water, simultaneously explaining the phenomenon of cooling, diffusion and the direction of time. Shannon defines

$$\text{information} = -\text{entropy}.$$

We may normalize the information by adding 1 to avoid negatives. We have

$$\text{information} = 1 - \text{entropy}$$

as used in this book. Classically, given a discrete random variable X, i.e., one whose range $R = (x_1, x_2, \ldots)$ is finite or countable. Let $p_i = P(X = x_i) = $ the probability for x_i to happen. The *entropy* $H(X)$ of X is defined by

$$H(X) = \sum_{p_i > 0} p_i \log_2 \frac{1}{p_i} = - \sum_{p_i > 0} p_i \log_2(p_i).$$

Let us take a simple example. Suppose we check one out of two boxes (e.g., male and female) to provide some information. Before choosing a box, both boxes have equal chance of being selected, and we only pick one. So, before the selection, the entropy is

$$\frac{1}{2} \log_2 2 + \frac{1}{2} \log_2 2 = 1 = 1 - \text{information}.$$

Therefore, the information is 0. After the selection, the entropy is

$$-1 \log_2 1 = 0 = 1 - \text{information}.$$

Therefore, the information is 1. The information gained is

$$\text{information} = 1 - 0 = 1.$$

For four boxes to be checked, for these requirements to be fulfilled, we either use the definition of entropy directly, or group them into two subsets, {box 1, box 2} and {box 3, box 4 }, then we pick one subset out of the two subsets and one box from the subset of two boxes, so

$$\text{information} = 1 + 1 = 2.$$

The material world tends to homogenize distributions, for instance, air tends to mix all components uniformly. These are the results of the increase of entropy. On the other hand, living beings tend to select from a mixed things to use some components only. These separations of components correspond to the increase of information. The natural world tends to increase the entropy and decrease the information. The living beings tends to increase the information and decrease the entropy. An ancient Chinese philosopher Lao Tzu (the founder of Taoism, cf. *Tao Te Ching*, Chapter 77) [2] said: "The heavenly Tao (i.e., the way of material world) takes from those who have too much, and it gives to those who have little or nothing (allude a flat plain is made of the erosion of maintain tops and filling of the valley bottoms). Ah, but the human way is different. Even the wealthiest leech the poor so they can have even more." Later on, the second part of his phrase was repeated in Matthew's Gospel [1] 12:13, 25:29 and known as "Matthew effect" in modern sociology. In modern times, Lao Tzu's words can be rephrased as "the way of nature is to increase entropy, the way of living beings is to increase information (i.e., to decrease entropy)." According to the second law of thermodynamics, the amount of entropy will increase or the amount of information will decrease, as it passes through noisy channels, *time*. Hamming code is a way of protecting parts of the information so they decrease less with the help of a proper decoding procedure. The meaning of Hamming's discovery can be better understood using the framework of Shannon's information theory. Hamming's discovery led to BCH codes, Reed–Solomon codes (which are widely used today), and geometric Goppa codes. Similar to how a tall tree starts from a tiny seed, the study of self-correcting codes eventually leads to many interesting problems for algebraic-geometry enthusiasts.

Let us return to the codes of life, DNA and RNA. Today, they are broken down into basic units called *genes*. There is an important phenomenon called *mutation* which is caused by environmental cosmic rays (lately, biologists add *free radicals* as other mutation agents) and is a driving force behind evolution. The cosmic rays input energy and change the configuration of the molecules of the genes. The mutations can be considered as errors. Some mutations are detrimental and are not welcomed by living beings. How does the body deal with these mutations? There may be some other machination, say a self-correcting function, to prevent those mutations from happening. We see that some genes are functional and some are *non-coding genes* or *junk genes*, and genetic engineers typically transplant only the functional genes to other species. We know that life is very economical, so why is it that the non-coding genes are passed down through the generations? Perhaps they exist for a self-correcting purpose, similar to the additional symbols $b_1 b_2 b_3$ in our example of Hamming code.

It turns out that DNA has some *proof-reading* capabilities which RNA lacks, although how those *proof-reading* capabilities function is unclear. These capabilities slow down the rates of mutations considerably. The fact that the lack of proof-reading capabilities for RNA make some viruses evolve rapidly and this phenomenon causes many problems for an individual's health and even a worldwide pandemic problem. The transplant of a single useful gene without the associate *proof-reading* capability might likely be dangerous. The self-correcting codes that occur in nature might be better than all of our algebraic coding theory.

Similar to the codes of life, civilizations and cultures themselves may be viewed as the transmission of codes through time. Due to the nature of decaying phenomena caused by historical events, thermodynamics, and cosmic rays, we may view the channel of time as a noisy channel. In his old age, Leonardo da Vinci worried about the decaying of his masterpieces. Preservation of our heritage becomes an important topic. One way might be using the self-correcting codes to prolong the useful period of our civilization and culture. Oral and written languages are important parts of heritages. In all oral and written languages, there are many *non-functional* parts which serve as check symbols. It is dangerous to delete these parts.

We live in the age of technology. Messages are transmitted in sequences of 0's and 1's through space. It is possible to make an error with noisy channels, and so self-correcting codes become vital to eradicate all errors (as long as the number of errors is small). Self-correcting codes are widely used in industries for a variety of applications including e-mail, telephone, remote sensing (e.g., photographs of Mars), CD, etc. We will present some essentials of the theory in this book.

Using linear algebra, we have the salient Hamming codes. The next level of coding theory is through the usage of ring theory, especially polynomials, rational functions, and power series to produce BCH codes, Reed–Solomon codes, and classical Goppa codes. The more advanced level of coding theory is an application of algebraic geometry to geometric Goppa codes. The aim of this book is to gradually bring interested readers to the most advanced level of coding theory.

PART I

Vector Space Codes

Chapter 1

Linear Codes

1.1. Real Curve Codes

In this chapter, we lay the foundation for coding theory using linear algebra, although certainly there are many other ways. For instance, we could simply send multiple copies of the same message and determine the correct one bit by bit by a majority vote, which is called *repetition* code. Alternatively, we could use real curve theory to construct a "self-correcting code" as follows.

Given data a_0, a_1, let us consider the line defined by linear equation

$$y = f(x) = a_0 x + a_1.$$

We transmitted the values $y_0 = f(0), y_1 = f(1)$ (and in general $y_i = f(i)$) instead of $\{a_0, a_1\}$, making the observation that $\{a_0, a_1\}$ and $\{y_0, y_1\}$ determine each other. There is no way to tell if the transmitted $\{y_0, y_1\}$ contain an error. To detect errors, assume that there is at most one error; we may transmit a group of three data, (y_0, y_1, y_2). If $(0, y_0), (1, y_1)$, and $(2, y_2)$ are on a line $L_3 \subset A_{\mathbb{R}}^2$, then there is no mistake since we assume that there is at most one error. If they are not on a line, then there is a mistake. However, we cannot decide which one is an error. To correct one possible error, we should add two more symbols $y_2 = f(2), y_3 = f(3)$ instead of just one more y_2 and transmit $\{y_0, y_1, y_2, y_3\}$ because we assume that there is at most one error, then there must be at least three correct values. For any three correct values $(f(i), f(j), f(k))$ among the four values, the corresponding points $(i, f(i)), (j, f(j)), (k, f(k))$ will lie on a line $L_4 \subset A_{\mathbb{R}}^2$, and hence the remaining one point $(\ell, f(\ell))$ is determined, i.e., since ℓ is correct, we just need to determine $f(\ell)$. That is, a brute-force search for

3

the correct triple will reveal which three values are consistent (for a line) and hence determine the extra one.

Let us consider the problem of correcting two errors. Let us add one more data $f(4)$. Now, we have $f(0), f(1), f(2), f(3)$, and $f(4)$. Can we correct two errors? Say, $(0, f(0)), (1, f(1)), (2, f(2))$ lie on a line. However, $(2, f(2)), (3, f(3)), (4, f(4))$ may lie on a different line. Then, we cannot tell which line is the correct one. We cannot correct the mistakes. We may modify the method in the previous paragraph to correct two errors. We shall add two more points $\{f(4), f(5)\}$ and transmit $\{y_0, y_1, \ldots, y_4, y_5\}$ because we assume that there are at most two errors; therefore, there are at least four correct values. Furthermore, any four correct values will determine the line and hence the remaining two values. That is, a brute-force search for the correct four tuple will reveal which four values are consistent (for a line). Thus, two errors can be corrected this way.

It is easy to generalize the above method to correct any number of errors.

Instead of lines, we may use curves of higher degrees. We may consider all quadratic curves. A quadratic curve is defined by the equation

$$y = f(x) = a_0 x^2 + a_1 x + a_2,$$

representing the original data $\{a_0, a_1, a_2\}$. We transmit the values $y_0 = f(0), y_1 = f(1)$, and $y_2 = f(2)$, instead of $\{a_0, a_1, a_2\}$, making the observation that $\{a_0, a_1, a_2\}$ and $\{y_0, y_1, y_2\}$ determine each other. There is no way to tell if the transmitted $\{y_0, y_1, y_2\}$ contain an error. To correct one possible error, we should add two more symbols $y_3 = f(3), y_4 = f(4)$ and transmit $\{y_0, y_1, y_2, y_3, y_4\}$. Suppose on the other end, the receiver receives $\{y_0', y_1', y_2', y_3', y_4'\}$ with one possible error. The receiver then determines which of the four tuples are consistent (for a quadratic curve) and then uses it to correct the fifth one. In general, we feel that if we consider curves of higher degrees and if we transmit sufficiently more points than necessary, say s more points, then we may correct $\lfloor \frac{s}{2} \rfloor$ errors. However, a brute-force search for the correct tuple will be time-consuming. We may modify this curve code slightly to produce the Reed–Solomon code (see Section 3.3) with a fast decoding process.

Both *repetition* codes and real curve codes are time-consuming for decoding. Instead, we focus on *linear codes*, which are more efficient and have fast decoding methods. We follow the historical development of the theory of self-correcting codes, primarily using techniques from linear algebra.

1.2. Preliminaries

1.2.1. *Groups and Fields*

We study the foundations of *algebra*. A kernel is *group theory*. Abstract *group theory* was an invention of Galois. It turns out to be an important concept in mathematics, physics, etc. Let us give the usual definition of abstract *group*.

Definition 1.1. Let G be a non-empty set. A binary operation $\cdot$ is a rule to assign an element $c = a \cdot b$, given any ordered pair of elements (a, b) in G. If $a \cdot b \in G$ for all $a, b \in G$, then we say that G is *closed* under the binary operation $\cdot$. If G is closed under $\cdot$ and has the following three properties, then we say $(G, \cdot)$ forms a *group* under the binary operation $\cdot$:

(1) *Associative law:* $(a \cdot b) \cdot c = a \cdot (b \cdot c)$.
(2) *The existence of identity:* there exists an element e such that $a \cdot e = e \cdot a = a$ for all $a \in G$.
(3) *The existence of inverse:* for any element a, there exists an element b such that $a \cdot b = b \cdot a = e$.

Sometimes, we omit the mention of $\cdot$ if it is obvious, and we simply say that G is a group. If $a \cdot b = b \cdot a$ holds for all $a, b \in G$ always, we say that G is an *abelian group* or *commutative group*. ∎

In primary and high schools, we have studied the *associative law*, *commutative law*, and *distributive law* for integers $\mathbb{Z}$, and we may have noticed that the additive identity $0 \neq 1$, the multiplicative identity. Now, we assume that the same rules are satisfied by the two imposed operations, addition and multiplication, on the set of symbols. Furthermore, we include one more useful rule which is valid for the set of rational numbers $\mathbb{Q}$, i.e., every non-zero element will have an inverse. We thus call a non-empty set **K** with two operations that satisfies all the preceding requirements a *field*. So, algebraically, a *field* behaves as the rational numbers. We use sets which are similar to the set of rational numbers and call them *fields*. Usually, we work on a set of symbols $\{a_1, \ldots, a_n\}$. We take our symbols from a finite field F_q (see the following), and we call them *letters* in our discussions of coding theory. A *field* $(\mathbf{K}, +, \cdot)$ is defined as follows.

Definition 1.2. Let **K** be a set with two operations $(+, \cdot)$, where $+, \cdot$ are binary operations between elements in **K** such that $a, b, c \in \mathbf{K}$ satisfies the following conditions:

(1) $(K, +)$ is an abelian group, with the identity denoted by 0;
(2) $(K\backslash\{0\}, \cdot)$ is an abelian group, with the identity denoted by 1;
(3) *Distributive law:* $(a + b) \cdot c = a \cdot c + b \cdot c$.

Then, we say that $(\mathbf{K}, +, \cdot)$ is a field. Sometimes, we omit the mention of $+, \cdot$ if it is obvious, and we simply state that $\mathbf{K}$ is a field.   ∎

Example 1: The well-known rational numbers $\mathbb{Q}$, real numbers $\mathbb{R}$, and the complex numbers $\mathbb{C}$ are all fields. There are other important fields $\mathbb{Z}/p\mathbb{Z}$, the field of the equivalent classes $[a]_p$ modulus a prime number p:

$$[a]_p = \{\text{all integers b such that } p \mid a - b\},$$

where p is a prime integer, and the addition and product are defined as

$$[a]_p + [b]_p = [a + b]_p, \ [a]_p \cdot [b]_p = [ab]_p.$$

It is routine to show that $\mathbb{F}_p = \mathbb{Z}/p\mathbb{Z}$ is a field if p is a prime number (see Exercises).   ∎

We have the following definition.

Definition 1.3. Let $\mathbf{K}$ be a field. Consider the set of repeatedly adding 1 as $S = \{1, 1 + 1, \ldots, \overbrace{1 + 1 + \cdots + 1}^{n}, \ldots\}$, where 1 is the multiplicative identity in the field $\mathbf{K}$. Then, either the set S does not contain 0 or the first n, such that $n \cdot 1 = 0$ must be a prime number p. In the first case, we say the field $\mathbf{K}$ has *characteristic* 0, and in the second case, we say the field has *characteristic* p.   ∎

The only thing we have to establish in Definition 1.3 is that with n, the smallest positive integer such that $n \cdot 1 = 0$, if n exists, then n must be a prime number. Otherwise, let $n = \ell \cdot m$ with $0 < \ell, m < n$, it follows from

$$n \cdot 1 = \ell \cdot m \cdot 1 = (\ell \cdot 1)(m \cdot 1) = 0.$$

Since we have a field, either $\ell \cdot 1 = 0$ or $m \cdot 1 = 0$. Therefore, n is not the smallest. A contradiction, i.e., ℓ is prime.

We need some basic knowledge of field theory. The reader is referred to *Commutative Algebra*, Vol. I, p. 60, by Zariski and Samuel [16], for field theory and the following corollary.

Corollary 1.4. *If $\mathbf{L}$ is an overfield of $\mathbf{K}$ and $x \in \mathbf{L}$, then x is algebraic over $\mathbf{K}$ if and only if $\mathbf{K}(x)$ is a finite extension of $\mathbf{K}$. In that case, if $n = [\mathbf{K}(x) : \mathbf{K}]$ (i.e., $n = $ the vector space dimension of $\mathbf{K}(x)$ over $\mathbf{K}$), then the degree of the minimal polynomial of x in $\mathbf{K}[X]$ is n.*

Proof. See the reference. □

The reader is referred to *Commutative Algebra*, Vol. I, p. 106, by Zariski and Samuel [16], for the following theorem.

Theorem 1.5. *If* **K** *is a field, then there exists an algebraic closure of* **K** *and any two algebraic closures of* **K** *are* **K**-*isomorphic fields.*

Proof. See the reference. □

1.2.2. *Finite Fields*

In *coding theory*, we are interested in a finite field **L**. Naturally, it contains some prime field $\mathbb{Z}_p$ which is of exactly p elements. Therefore, **L** is a finite-dimensional vector space over $\mathbb{Z}_p$. By the vector space consideration of dimension, let the dimension be n, then putting a coordinate system to it, we conclude that it has precisely p^n elements, using the following theorem.

Theorem 1.6. *If* **K** *is a finite field, then there are exactly* $p^n = q$ *elements in* **K** *for some suitable prime number* p *and a positive integer* n. *If we fix an algebraic closure* Ω *and only consider subfields of* Ω, *then the field of* p^n *elements exists uniquely and will be called* $\mathbf{F_{p^n}} = \mathbf{F_q}$. *We may use either* $\mathbb{Z}_p$ *or* $\mathbf{F_p}$ *to denote a finite field of* p *elements.*

Proof. See the above references. □

1.2.3. *Vector Spaces*

Using letters, we form messages (which maybe ordinary words and sentences and articles). Note that some letters may indicate empty space or empty line. Given any positive integer k (i.e., the size of a page), we may chop a long article into many blocks of the same length k, just as we chop a long book into many pages, and pad the last block. We shall only work on a single block of size k. Note that messages (which may be ordinary words, sentences, and articles) are more than simple letters. If there is no relation among letters for one message, then it is hard to locate and correct errors. It is to our advantage to introduce *algebraic* relations for letters in a message. We shall put the simplest relation, the structure of a *vector space* (see later), among letters in a single block of length k to help us locate and correct errors. We will make a detailed study of the coding theory thus discussed.

Before we go further, we need some elementary knowledge of *Linear Algebra.* Let us define the concept of *vector space.*

Definition 1.7. Let $(\mathbf{K}, +, \cdot)$ be a field and $\mathbf{V}$ be a non-empty set. A *vector space* $(\mathbf{V}, +, \cdot, \mathbf{K})$, where $+$ is a binary operation between elements in $\mathbf{V}$ and $\cdot : \mathbf{K} \times \mathbf{V} \mapsto \mathbf{V}$ is a binary operation with $a, b \in \mathbf{K}$ and $v, u \in \mathbf{V}$ satisfying the following conditions:

(1) $(\mathbf{V}, +)$ is an abelian group.
(2) *Associative law:* $(a \cdot (b \cdot v)) = (a \cdot b) \cdot v$, for all a, b, v.
(3) *Distributive law:* $(a + b) \cdot v = a \cdot v + b \cdot v$ and $a \cdot (v + u) = a \cdot v + a \cdot u$.
(4) $1 \cdot v = v$.

Sometimes, we omit the mention of $(\mathbf{V}, +, \cdot)$ if they are obvious and say $\mathbf{V}$ is a vector space. ∎

A set of vectors $\{v_i\}_{i \in I}$ is called a set of generating vectors if for any vector $v \in \mathbf{V}$, we always have an expression $v = \sum_{\text{finite}} a_i v_i$. A set of vectors $\{v_i\}_{i \in I}$ is called a set of linearly independent vectors if for any expression $0 = \sum_{\text{finite}} a_i v_i$, we must have $a_i = 0 \ \forall i$. A set of vectors $\{v_i\}_{i \in I}$ is called a basis of $\mathbf{V}$ if it is a set of generating and linearly independent vectors.

A common theorem of the vector spaces is that for a given vector space $\mathbf{V}$, all bases will have the same cardinality, which is called the *dimension* of the vector space $\mathbf{V}$. Any vector space is said to be finite-dimensional if it has a finite basis. In coding theory, we only use finite-dimensional vector spaces.

If we have two fields $\mathbf{L} \supset \mathbf{K}$, then clearly $\mathbf{L}$ is a vector space over $\mathbf{K}$.

1.2.4. *Matrices*

Consider a matrix A. The matrix A is said to be in a *reduced row echelon form* if it satisfies the following conditions: (1) All rows are of the form $[0, \ldots, 0, 1, \ldots]$, where the first non-zero term is 1 and happens at n_jth position. Usually, this term is called the *pivot term.* (2) If $i < j$, then $n_i < n_j$. (3) On the n_jth column, only that particular coefficient is 1, while all other coefficients are zeroes. It follows from a well-known theorem that by *row operations*, any matrix can be reduced to its *reduced row echelon form*.

1.3. Vector Space Codes

In general, setting an algebraic closure Ω of $\mathbb{Z}_p$, we will consider a finite field $\mathbf{K}$ of $q = p^m$ elements between Ω and $\mathbb{Z}_p$. The field is uniquely determined by the number q and will be denoted by $\mathbf{F}_q$. The detailed treatments of finite fields is given in Section 2.2.

Let us refer to the finite field $\mathbf{F}_q$ as our set of *letters*, then any message $a_1 a_2 \cdots a_k$ can be considered as a horizontal vector $[a_1 a_2 \cdots a_k]$ in the vector space $\mathbf{F}_q^k$.

A basis $\{e_1, \ldots, e_k\}$ is called a standard basis if any element e_i is of the form $e_i = (0, \ldots, 0, 1, 0, \ldots, 0)$. For the purpose of coding theory, we will only consider the standard basis in $\mathbf{F}_q^k$ since a message is composed of letters with positions, and a letter will be denoted by an element $a_i \in \mathbf{F}_q$, while a message will be denoted by $\sum_i a_i e_i$ in the standard basis. Since we use horizontal vectors, all matrices of linear action will act from the right. The theory of Gaussian reduced row echelon forms are significant, and we shall allow a permutation of columns (which corresponds to exchanging the positions of letters). The vector space $\mathbf{F}_q^k$ with the *standard basis* will be called the *message space* $\mathbf{V}$.

Let $n \geq k$. We should consider another vector space $\mathbf{F_q^n} = \mathbf{U}$ as the *word space*, and any vector in $\mathbf{F_q^n} = \mathbf{U}$ will be called a *word*. We shall repeat our construction of standard basis, and a word will be any vector $\sum_i^n a_i e_i$, and we shall allow permutation of columns. A system of linear equations defines a subspace $\mathbf{C}$ of dimension k in $\mathbf{F_q^n}(= \mathbf{U})$. Any element in $\mathbf{C}$ will be called a *code word*. The subspace $\mathbf{C}$ will be referred to as the *code space*.

Definition 1.8. An $[n, k]$ code is a k-dimensional subspace $\mathbf{C}$, which is called *code space* of an n-dimensional vector space $\mathbf{F}_q^n (= \mathbf{U})$, which is called *word space*, with standard basis. Sometimes, the number k will be referred to as the *rank* of $\mathbf{C}$. A *coding* is a one-to-one (i.e., injective) map from the message space $\mathbf{V}$, which is another copy of vector space $\mathbf{F_q^k}$, to the code space $\mathbf{C}$.

All concepts and theorems in coding theory are independent of the ordering of the basis vectors, so we will allow permutations of the standard bases in the message spaces and word spaces. Two $[n, k]$ codes $\mathbf{C_1}, \mathbf{C_2}$ are said to be *equivalent* if their code spaces $\mathbf{C_1}, \mathbf{C_2}$ differ by permutations of the standard bases of the message space and word space. ∎

A k-dimensional subspace $\mathbf{C}$ is generally defined by $n - k$ equations. To apply Gaussian elimination to simplify the system, we are restricted to using row operations and permutations of columns only. Therefore, we may assume that the system of equations is of the following *standard* form:

$$y_1 \quad = c_{1,1}x_1 + \cdots + c_{1,k}x_k,$$

$$\cdots \tag{2}$$

$$y_{n-k} = c_{n-k,1}x_1 + \cdots + c_{n-k,k}x_k.$$

It follows that the *generator matrix* G can be written in the following *standard* form:

$$G = \begin{pmatrix} 1 & 0 & \cdots & 0 & c_{1,1} & \cdots & c_{n-k,1} \\ 0 & \cdots & \cdots & \cdots & \cdots & \cdots & \cdots \\ \cdots & \cdots & \cdots & 0 & \cdots & \cdots & \cdots \\ 0 & \cdots & 0 & 1 & c_{1,k} & \cdots & c_{n-k,k} \end{pmatrix},$$

with $[x_1 \cdots x_k] \times G = [x_1 \cdots x_k y_1 \cdots y_{n-k}]$. The symbols x_i will be called *message symbols* and the symbols y_i will be called *check symbols*. Define G' using equation (2), where

$$G = \begin{pmatrix} I & G' \end{pmatrix}$$

and I is the $k \times k$ identity matrix.

Observe that if we rewrite the system of equations in equation (2) in the form

$$y_1 \quad - c_{1,1}x_1 - \cdots - c_{1,k}x_k = 0,$$

$$\cdots \tag{2'}$$

$$y_{n-k} - c_{n-k,1}x_1 - \cdots - c_{n-k,k}x_k = 0,$$

then the *check matrix* H can be written as

$$H = \begin{pmatrix} -c_{1,1} & \cdots & -c_{n-k,1} \\ \cdots & \cdots & \cdots \\ \cdots & \cdots & \cdots \\ -c_{1,k} & \cdots & -c_{n-k,k} \\ 1 & \cdots & 0 \\ \cdots & \cdots & \cdots \\ 0 & \cdots & 1 \end{pmatrix}.$$

It is easy to verify that $G \times H = 0$, hence we have $[x_1 \cdots x_k] \times G \times H = [0 \cdots 0]$. We may write

$$H = \begin{pmatrix} -G' \\ J \end{pmatrix},$$

where J is the $(n - k) \times (n - k)$ identity matrix.

For a longer message, we chop it into blocks, which are vectors in $\mathbf{F}_q^k$, padding the end if necessary and applying the above method for each block. This is the so-called *block* code.

We may identify the message space $\mathbf{V}$ with a subspace $\{[x_1 \cdots x_k 0 \cdots 0] : x_i \in \mathbf{F}_q\}$. If there is no confusion, we will call this subspace the message space $\mathbf{V}$.

Let us consider the coset space $\mathbf{U}/\mathbf{V}$ the image space of the mapping defined by H. It is called the *syndrome space*. It can be canonically identified with the subspace $\mathbf{W} = \{[0 \cdots 0 y_1 \cdots y_{n-k}] : y_j \in \mathbf{F}_q\}$. For any element $a \in \mathbf{U}$, $a \times H$ is called the *syndrome* of a. We have the following proposition.

Proposition 1.9. *Let $a, b \in \mathbf{U}$. Then, they have the same syndrome or they belong to the same coset iff $a \times H = b \times H$. In particular, a is in the code space iff $a \times H = 0$.* ∎

Example 2: The example of Hamming in the introduction is a $[7, 4]$ code. Its syndrome space $\mathbf{W}$ is $\{[0000 y_1 y_2 y_3] : y_j \in \mathbf{F}_q\}$. Let us consider a general $[n, k]$ code $\mathbf{C}$. From the check matrix H, we may add $n \times k$ zeroes to form the following $n \times n$ matrix $\bar{H}$:

$$\bar{H} = \begin{pmatrix} 0 & \cdots & 0 & -c_{1,1} & \cdots & -c_{n-k,1} \\ 0 & \cdots & 0 & \cdots & \cdots & \cdots \\ 0 & \cdots & 0 & \cdots & \cdots & \cdots \\ 0 & \cdots & 0 & -c_{k,1} & \cdots & -c_{n-k,k} \\ 0 & \cdots & 0 & 1 & \cdots & 0 \\ 0 & \cdots & 0 & \cdots & \cdots & \cdots \\ 0 & \cdots & 0 & 0 & \cdots & 1 \end{pmatrix}.$$

Then, it is easy to see that $\bar{H}^2 = \bar{H}$, and $\bar{H} : a \mapsto a \times \bar{H}$ is the projection of the whole space $\mathbf{U}$ to the syndrome space $\mathbf{W}$. ∎

1.4. Distances

The only natural metric in $\mathbf{F}_q$ is the discrete one, i.e., we have $d(a,b) = 1$ if $a \neq b$ and $d(a,b) = 0$ if $a = b$. We generalize this distance function to an n-dimensional vector space $\mathbf{F_q^n}(= \mathbf{U})$ by using the following definitions.

Definition 1.10. Let $a = [a_1 \cdots a_n], b = [b_1 \cdots b_n] \in \mathbf{U}$. Then, we define the *Hamming distance* between a and b as the product distance of the above-mentioned natural metric in $\mathbf{F}_q$ (which may be called a sum distance), i.e., $d(a,b) = \sum d(a_i, b_i)$, where $d(a_i, b_i)$ is the natural distance of elements in the field $\mathbf{F}_q$. ∎

Definition 1.11. Let $a = [a_1 \cdots a_n] \in \mathbf{U}$. Then, we define the *Hamming weight* of a as $w(a) = d(a,0)$. ∎

We have the following natural theorem.

Theorem 1.12. *The Hamming distance is a distance, i.e.,*

$$(1): d(a,a) = 0,$$

$$(2): d(a,b) = d(b,a),$$

$$(3): d(a,b) + d(b,c) \geq d(a,c).$$

Proof. It is routine for all product distances. □

Proposition 1.13. *We always have* $d(a,b) = w(a-b) = d(a-b,0)$.

Proof. It is evident. □

Definition 1.14. The minimal distance of a set S is $\min\{d(a,b) : a \neq b, \ a,b \in S\}$. Sometimes, the minimal distance of a set S is called the *Hamming distance* of S. ∎

1.5. Maximum Likelihood Decoding

From the above discussion, we know that a coding theory is determined by its code space, which is a subspace of the word space. In this section, we show that given any subspace of a finite-dimensional vector space, a coding theory will be determined naturally. This means that we will show that the coding theory corresponding to the theory of subspaces of a finite-dimensional vector space.

Let a subspace $\mathbf{C}$ of a finite-dimensional space $\mathbf{U}$ be given. Let us call $\mathbf{C}$ the code space and $\mathbf{U}$ the word space. A finite-dimensional subspace $\mathbf{C}$ is determined by a finite system of equations, and by Gaussian row operations and column permutations, we may assume it to be of the following forms:

$$y_1 = c_{1,1}x_1 + \cdots + c_{1,k}x_k,$$

$$\cdots \tag{2''}$$

$$y_{n-k} = c_{n-k,1}x_1 + \cdots + c_{n-k,k}x_k.$$

It is easy to see that all $\{x_1, \ldots, x_k\}$ define a message space $\mathbf{V}$. Let us only transmit vector v from $\mathbf{C}$. Suppose we have received a message $v' = [a'_1 \cdots a'_n]$, and we know the code word $v = [a_1 \cdots a_n]$ is in the code space $\mathbf{C}$, but perhaps $v' \notin \mathbf{C}$. How do we recover the code word v? Let us assume that $v' = v + e$, where e is the *error* vector. Since $v = v' - e$, v and e uniquely determine each other with v' known. However, with only v' known, we cannot uniquely determine e. We may select possible e from the syndrome space with $e \times \bar{H} = v' \times \bar{H}$, where $\bar{H}$ is used in Example 2 of the preceding section, and let $v = v' - e$. Note that we have

$$v \times \bar{H} = (v' - e) \times \bar{H} = 0.$$

Therefore, we conclude that $v \in \mathbf{C}$, the code space. However, upon closer examination, we discover that we may use any element in the same coset as e. Up to now, there is no particular reason to pick any one from the syndrome space. We have to find some criterion for the *best possible* selection.

Let us consider the real situation. Let us fix a position for the error vector e; a zero coordinate indicates that there is no error in that position, whereas a non-zero coordinate indicates that the letter at that position should be corrected. Let p_i be the probability (which is due to the machine or the program and is independent of v') that the error vector e at a particular position has a value i. For simplicity, we assume that the following definition is satisfied.

Definition 1.15 (Symmetric channel). If $p_i = p_j(= p)$ for all $i, j \neq 0$, and $p_0 > 1/2$, then the channel is said to be a *symmetric channel*. ∎

Let q be the number of symbols i, j. Note that $\sum_{i=0}^{q-1} p_i = 1$; therefore, $p = p_i < 1/(2(q-1))$ for all $i \neq 0$. Note that p_0 is the probability that there is no error at this particular position. The condition $p_0 > 1/2$ simply means that it is less likely to have an error. This is reasonable. We shall assume that we have a *symmetric channel.*

Proposition 1.16. *For a symmetric channel, the probability $P(e)$ for an error vector e to happen is $p^{w(e)}(1 - w(e)p)^{q-w(e)}$, where $w(e)$ is the Hamming weight of e. Therefore, $P(e) > P(e')$ iff $w(e) < w(e')$.*

Proof. It is evident.   □

Let us continue our discussion of taking a subspace **C** as code space; an *ad hoc* decoding method is the following *maximum likelihood decoding* method. The *maximum likelihood decoding* is to find an error vector e in the coset S of v' such that

$$w(e) = d(e,0) = \min\{w(u) : \ u \in S\}.$$

It means that the *maximum likelihood decoding* is to select the one correction with the least amount of corrections. The previous proposition explains the meaning of the term *maximum likelihood decoding*. We further define the following.

Definition 1.17. A *coset leader* e of a coset S is an element e in S such that

$$w(e) = d(e,0) = \min\{w(u) : \ u \in S\}. \qquad ■$$

See Figure 1.1.

In the case of real numbers, the *maximum likelihood decoding* is to orthogonally project the vector v' to the subspace **C**. Usually, the orthogonal projection of v' is unique. Figure 1.1 over the field of real numbers is misleading in the context of a finite field, where the coset leaders

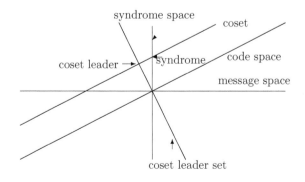

Figure 1.1.   Various spaces.

of a vector v' only forms a set rather than a point as in the real case. For instance, we have the following example.

Example 3: Let us consider a $[6, 3]$ code $\mathbf{C}$ over the prime field $\mathbf{F_2} = \mathbb{Z}/2\mathbb{Z}$ with the following check matrix H:

$$H = \begin{pmatrix} 1 & 1 & 0 \\ 0 & 1 & 1 \\ 1 & 0 & 1 \\ 1 & 0 & 0 \\ 0 & 1 & 0 \\ 0 & 0 & 1 \end{pmatrix}.$$

Furthermore, we let $\bar{H}$ be the following matrix:

$$\bar{H} = \begin{pmatrix} 0 & 0 & 0 & 1 & 1 & 0 \\ 0 & 0 & 0 & 0 & 1 & 1 \\ 0 & 0 & 0 & 1 & 0 & 1 \\ 0 & 0 & 0 & 1 & 0 & 0 \\ 0 & 0 & 0 & 0 & 1 & 0 \\ 0 & 0 & 0 & 0 & 0 & 1 \end{pmatrix}.$$

Let the coset S be with syndrome $[000111]$, i.e., $s \times \bar{H} = [000111]$ for any $s \in S$; then, it is easy to see that no element s with $w(s) = 1$ is in S, while there are three elements $[100001], [010100]$, and $[001010]$ in S with $w(s) = 2$. They are all coset leaders, and they are not unique. ∎

If there are several coset leaders for a coset, we will identify any one of them as the coset leader. The maximum likelihood decoding procedure is as follows: For any message v', we find the coset S where it lies (usually by finding the syndrome of v' which determines the coset S); then, we find a coset leader e which is the most likely error vector in S. Finally, we correct the error by taking $v = v' - e$. In engineering, we may have a table of $\{(syndrome, coset\ leader)\}$ for the sake of convenience. Since the syndrome space is of dimension $n - k$, it consists of q^{n-k} elements. If $n - k$ is small, we may precompute the table and decode accordingly.

Note that this particular decoding procedure may not be effective for all codes, and it may decode to a wrong word, so we shall look for other possible decoding procedures. The advantage of this procedure is that it does exist for any code. Therefore, we may study any code with the decoding procedure of maximum likelihood decoding, which may not be effective. In all cases, the *maximum likelihood decoding is an ad hoc method of decoding.* The advantage is that on using it, every subspace $\mathbf{C}$ establishes a self-correcting code. Therefore, there is a one-to-one correspondence between the set of self-correcting codes and the set of all subspaces.

Let us continue our discussion on possible codes.

1.6. Dual Codes

Let $\mathbf{C}$ be a linear code, i.e., $\mathbf{C}$ is a vector subspace of $\mathbf{F}_q^n$. Let us introduce a *pairing* $\langle v \mid u \rangle$ on $\mathbf{F}_q^n$ as follows.

Definition 1.18. Let $v = (v_1, \ldots, v_n)$ and $u = (u_1, \ldots, u_n)$ be two vectors $\in \mathbf{F}_q^n$ with respect to a fixed basis. Note that we only allow a basis to be a permutation of the standard basis. We define a *pairing* $\langle v \mid u \rangle$ as

$$\langle v \mid u \rangle = \sum_{i=1}^{n} v_i u_i.$$

∎

Note that a pairing is not an inner product, since there is no sense of sizes in $\mathbf{F}_q^n$, we cannot say $\langle v \mid v \rangle > 0$ if $v \neq 0$. However, the *dual space* exists, and we define the following.

Definition 1.19. Let $\mathbf{C}^\perp = \{v \in F_q^n : \langle v \mid u \rangle = 0 \text{ for all } u \in \mathbf{C}\}$. If $\mathbf{C}$ is a linear code, then $\mathbf{C}^\perp$ is called the *dual code* of $\mathbf{C}$. ∎

Proposition 1.20. *Given $\mathbf{C}$ a linear code, its dual $\mathbf{C}^\perp$ is also a linear code. If $\mathbf{C}$ is an $[n, k]$ code, then $\mathbf{C}^\perp$ is an $[n, n - k]$ code.*

Proof. The proof is left to the reader. □

Exercises

(1) Prove that $\mathbb{Z}/p\mathbb{Z}$ is a field, while $\mathbb{Z}/p^m\mathbb{Z}$ is not a field for $m \geq 2$.

(2) Let us consider the repetition code $[a_1 \cdots a_n] \; :\mapsto \; [a_1 a_1 \cdots a_1 \cdots a_n a_n \cdots a_n]$, where each digit repeats itself m times. Find the generator matrix, check matrix, and the minimal distance of this code.

(3) Prove that for a given binary $[n, k]$ code with at least one word of odd weight, all code words of even weight form an $[n, k - 1]$ code.

(4) Let us consider the example of Exercise (2). If we want to correct two errors, how long should be the code? If we want to correct ℓ errors, how long should be the code?

(5) Prove that if $\mathbf{C}$ is an $[n, k]$ code, then $\mathbf{C}^{\perp}$ is an $[n, n - k]$ code.

(6) Let $\mathbf{C}$ be a code space. Show that
$$\min\{d(a, b) : a, b \in \mathbf{C}\} = \min\{w(a) : a \in \mathbf{C}\}.$$

(7) Let $\mathbf{C}$ be the *repetition* code with $n = 3 = m$ (cf. Exercise (2)). What is $\mathbf{C}^{\perp}$?

(8) Let us use the lines for a coding theory of plan to correct one error. We received $(1, 3, 3, 4)$. Assume that there is at most one error. Which of the four digits is an error, and which are the correct four digits?

1.7. Hamming Code

Let us consider the case of a *binary code*, i.e., $q = 2$, $\mathbf{F}_q = \mathbf{F}_2$. For $n \geq 3$, a *Hamming code* is a code with the check matrix H, in which the rows of H are all non-zero elements of $\mathbf{F}_2^n$. Note that the number of rows is $2^n - 1$, H is a $(2^n - 1) \times n$ matrix, and the code thus constructed is a $[2^n - 1, 2^n - n - 1]$ code. After a suitable permutation of rows, we may assume that the last n rows form an identity matrix J and hence H is in *standard form*.

For the purpose of decoding, we may arrange the rows of H differently. We will present elements in $\mathbf{F}_2^n$ as integers as follows. Recall the *binary expansion* of an integer i between 0 and $2^n - 1$ as follows:
$$i = \sum_{j=1}^{n} a_j 2^{j-1}, \quad 0 \leq i \leq 2^n - 1, \quad a_j = 0, 1,$$
and we put $[a_1 \cdots a_n]$, which is an element of $\mathbf{F}_2^n$, as the ith row of H. Now, suppose that a received word $[b_1' \cdots b_{2^n-1}']$ has at most one error. It means that if $[b_1 \cdots b_{2^n-1}]$ is the code word, then either
$$[b_1' \cdots b_{2^n-1}'] = [b_1 \cdots b_{2^n-1}]$$
or
$$[b_1' \cdots b_{2^n-1}'] = [b_1 \cdots b_{2^n-1}] + [0 \cdots 010 \cdots 0].$$
The computer calculates
$$[b_1' \cdots b_{2^n-1}'] \times H = [c_1 \cdots c_n].$$
If $[c_1 \cdots c_n] = [0 \cdots 0]$, then we can see there is no error because if there is a single error, we must have
$$[b_1' \cdots b_{2^n-1}'] \times H = ([b_1 \cdots b_{2^n-1}] + [0 \cdots 010 \cdots 0]) \times H$$
$$= [0 \cdots 010 \cdots 0] \times H \neq [0 \cdots 0],$$
which is a contradiction. If $[c_1 \cdots c_n] \neq [0 \cdots 0]$, then there must be an error, and by the above computation, it is easy to see that the error appears at

the ith position, where

$$i = \sum_{j=1}^{n} c_j 2^{j-1}.$$

The method is effective in both locating and correcting the error since error correction is simply switching 0 and 1. The shortcoming is that the Hamming codes are unable to correct several errors (in particular, a burst of errors).

The *Hamming code* is the grandfather of all self-correcting codes, it calls for more theoretical study of Hamming codes. We have the following proposition.

Proposition 1.21. *The minimal distance of all pairs of elements of any Hamming code* $\mathbf{C}$ *is three. It means that*

$$\min(w(a)) = 3, \text{ for all } a \neq 0 \text{ in } \mathbf{C}.$$

Proof. We have $a \in \mathbf{C} \Leftrightarrow a \times H = 0$. This equation can be interpreted as a linear equation among the rows of H, i.e., the minimal distance is the minimal number of row vectors which are linearly dependent. Since all row vectors of H are non-zeroes, we have $\min(w(a)) > 1$. Since any two row vectors are distinct, we have $\min(w(a)) > 2$. We can easily find two vectors which are added to a third row, so we conclude $\min(w(a)) = 3$. $\qquad \square$

The concept of the above proposition is very important. So, we have the following definition.

Definition 1.22. Let $\mathbf{C}$ be an $[n, k]$ code. If

$$d = \min(w(a)) \text{ for all } a \neq 0 \text{ in } \mathbf{C},$$

then $\mathbf{C}$ will be called an $[n, k, d]$ code. $\qquad \blacksquare$

The following proposition was discovered by Singleton in 1964.

Proposition 1.23 (Singleton bound). *Let* $\mathbf{C}$ *be an* $[n, k, d]$ *code. Then,* $d \leq n - k + 1$.

Proof. Let π be the projection of the word space to the last $n - d + 1$ coordinates. Let us restrict π to the code space $\mathbf{C}$. Let two code words $c \neq c'$ or $0 \neq c - c' \in \mathbf{C}$. Then, there must be at least d non-zero coordinates for $c - c'$. Clearly, if all the last $n - d + 1$ coordinates of $c - c'$ are zeroes, then the number of non-zero coordinates of $c - c' \leq n - (n - d + 1) = d - 1$. This is contradictory to the definition of d. Hence, $\pi(c - c') \neq 0$

or $\pi(c) \neq \pi(c')$. Therefore, the restriction of π is a one-to-one map on $\mathbf{C}$. Looking at the dimensions, we conclude that $k = \dim(\mathbf{C}) \leq n - d + 1$ and $d \leq n - k + 1$. $\qquad\square$

Definition 1.24. A code that satisfies the Singleton bound with equality, i.e., $d = n - k + 1$, is called an *MDS* code (maximum distance separable code). $\qquad\blacksquare$

Note that all Hamming codes are $[2^n - 1, 2^n - n - 1, 3]$ codes for $n \geq 3$. Note that for it to be an *MDS* code, we must have $3 = (2^n - 1) - (2^n - n - 1) + 1 = n + 1 \geq 4$, which is not true. Therefore, none of them is an *MDS* code. We have the following proposition.

Proposition 1.25. *Let $\mathbf{C}$ be an $[n, k, d]$ code. Then, for any received word $a' = [a'_1 \cdots a'_n]$ for a code word $a = [a_1 \cdots a_n]$ with less than or equal to $\lfloor (d - 1)/2 \rfloor$ errors, there is at most an element $a \in \mathbf{C}$ such that $d(a, a') \leq \lfloor (d - 1)/2 \rfloor$. Therefore, if the number of errors is restricted by $\lfloor (d - 1)/2 \rfloor$, then the decoded word is unique if it existed.*

Proof. It follows from the statement that there is a code word a such that $d(a, a') \leq \lfloor (d - 1)/2 \rfloor$. If there are two elements $a = [a_1 \cdots a_n]$, $b = [b_1 \cdots b_n] \in \mathbf{C}$ which satisfy the criteria of the proposition, then $d(a, b) \leq d(a, a') + d(b, a') < d$, which is a contradiction.

Furthermore, the number of errors allowed is the maximal of $d(a', a)$, where a is the code word and a' is the received word. The last statement is obvious. $\qquad\square$

Remark: In general, we may define (may be nonlinear) code as a subset M of a vector space $\mathbf{F}_q^n$. Let r be a received word which may be any element in $\mathbf{F}_q^n$ as $c \in M$ the original code word of r. Then there are two ways of decoding: (1) let us use the **maximum likelihood decoding** (cf. Section 1.5) to decode. The problem is that it is neither always correct nor efficient. (2) The following is what we will do in the coding theory. With the minimal distance d defined as

$$d = \min\{d(a, b) : a \neq b \in M\}.$$

Given an integer $t \leq \lfloor (d - 1)/2 \rfloor$, we return an *error message* when the number of errors is $d(r, M) > t$; otherwise, we have $d(r, M) \leq t$. Note that this implies that there is a unique element $c \in M$ such that $d(r, c) \leq t$, and we correct r to c. Then, we construct a decoder which corrects up to t errors. The important properties of the decoder are to determine quickly if

$d(r, M) > t$ and find quickly the correct c if $d(r, M) \leq t$. What we can do now is not up to expectation: if there are less than or equal to t errors, the decoder will find the correct c for us; if $d(c, M) > t$, then the decoder will find a $c' \in M$ such that $d(c', r) \leq r$ (with a small probability) and return an *error message* if c' cannot be found (with a large probability).   ∎

Remark: Let us consider the example of the beginning of Section 1.1. Let us consider the linear case. If we only consider $L_3 = \{(0, y_0), (1, y_1), (2, y_2)\}$, then it is easy to see that any two lines passing two points must pass the third one, while two lines can share a common point; therefore, $d = 2$ and $\lfloor (d - 1)/2 \rfloor = 0$. Thus, it cannot correct the error. If we consider $L_4 = \{(0, y_0), (1, y_1), (2, y_2), (3, y_3)\}$, then it is easy to see that any two lines passing two points must pass the third one, while two lines can share one point; therefore, $d = 3$ and $\lfloor (d - 1)/2 \rfloor = 1$. Thus, it can correct one error.   ∎

Note that according to the above proposition, Hamming code may correct $1 = \lfloor (3-1)/2 \rfloor$ error. This we already know. Later, we will construct $[n, k, d]$ codes for large d and decode more than one error.

Another important property of Hamming $[2^n - 1, 2^n - n - 1, 3]$ codes $\mathbf{C}$ is that any word $a' = [a_1' \cdots a_{2^n-1}']$ is within a distance of 1 of the code space because $a' \times H = [c_1, \ldots, c_n] = [0 \cdots 010 \cdots 0] \times H$; therefore, $(a' + e) \times H = [0 \cdots 0]$, where $e = [0 \cdots 010 \cdots 0]$ and $a' + e \in \mathbf{C}$. We define the following.

Definition 1.26. Let $\mathbf{C}$ be an $[n, k, d]$ code. If all words in the word space $\mathbf{F_q^n}$ are within a distance of $\lfloor (d-1)/2 \rfloor$ of $\mathbf{C}$, then $\mathbf{C}$ will be called a *perfect code*.   ∎

Definition 1.26 means that any word can be corrected with maximal $t = \lfloor (d - 1)/2 \rfloor$ errors to an unique code word.

Let us compute the probability of a decoder, which can correct t errors on an $[n, k]$ code, to fail. For instance, for Hamming codes $t = 1$, if there are $t + 1$ or more errors, then the Hamming code decoder will not decode correctly. Let the channel have a probability p of being incorrect and q of being correct. Then, $p + q = 1$ and

$$1 = 1^n = (p + q)^n = \sum_{i=0}^{n} \mathbf{C}_i^n p^i q^{n-i}.$$

The probability r of failing to decode or decoding improperly is

$$r = \sum_{i=t+1}^{n} \mathbf{C}_i^n p^i q^{n-i} = 1 - \sum_{i=0}^{t} \mathbf{C}_i^n p^i q^{n-i}.$$

Exercises

(1) Show that a Hamming code is a perfect code (note that in the definition of Hamming code, we assume that $n \geq 3$).

(2) Let $\mathbf{C}$ be a binary perfect code of length n with minimum distance 7. Show that $n = 7$ or $n = 23$.

(3) Let p be a prime number and $q = p^m$. A q-ary Hamming code of length $(q^n - 1)/(q - 1)$ is defined by a check matrix H with the following properties: (1) the zero vector is not any row vector, (2) any two row vectors are linearly independent, and (3) any non-zero vector is linearly dependent on one of the row vectors. Show that it is a perfect code.

(4) Set up a computer program to decode a Hamming code.

(5) Show that if there are more than two errors in a received word r for a Hamming code, then r will be decoded to a wrong code word.

1.8. Shannon's Theorem

Shannon's theorem is a guiding light of coding theory. For all the different codes, we need common standards of measurements to compare them. We sometimes define the common standards for linear code first and then generalize them for arbitrary codes. Let us consider the efficiency of codes first. We have the following definition.

Definition 1.27. The *rate of information* R of an $[n, k]$ code is defined to be k/n. ∎

Certainly, we have $0 < k/n \leq 1$, and we want the number k/n as large as possible. It is obvious that if $k/n = 1$, then the code cannot correct any error. Let us consider all codes (linear or otherwise). A *code* is defined as a subset $M = \{a_1, \ldots, a_m\}$, with m elements, of the word space $\mathbf{F}_q^n = \mathbf{U}$. We may use the maximum likelihood decoding to decode any received word a' to a, as $a \in M$ with $d(a, a')$ minimal (cf. Remark of Proposition 1.25). We shall generalize the above definition to the general cases. Let us use the linear codes as guidance. For a linear code space of dimension k, the number of elements is q^k, and $k/n = \log_q(q^k)/n$. Therefore, we naturally generalized the above definition to the following.

Definition 1.28. The *rate of information* R of a code M of m elements in F_q^n is defined to be

$$R(M) = n^{-1} \log_q m. \qquad \blacksquare$$

Let P_i be the probability of incorrect decoding for $a_i \in M$. Then,

$$P(M) = m^{-1} \sum P_i$$

is defined to be the average probability of incorrect decoding for the code M.

Definition 1.29. The δ *rate of distance* of an $[n, k, d]$ code is defined to be $\delta = d/n$. $\qquad \blacksquare$

Proposition 1.30. *The rate of information of an $[n, k, d]$ code is at most $1 + 1/n - \delta$.*

Proof. It follows from Proposition 1.23. $\qquad \square$

It follows from the preceding proposition that as the rate of distance δ gets larger for a linear code, the rate of information gets smaller. We are working towards the statement of Shannon's theorem [33], which has been very influential to the development of coding theory. We will consider all codes (linear or otherwise).

Suppose that the message is transmitted through a *binary symmetric channel*, i.e., the symbols 0 and 1 have equal probability $\wp$ of being transmitted incorrectly. The value of $\wp$ is $< 1/2$. We shall have a common measurement of all channels of this kind. There are only two possible outcomes, either correct or incorrect; the classical information function $\wp \log_2 \wp + (1 - \wp) \log_2 (1 - \wp)$ gives a good measure, with the exception that its values may be negative, so add 1 to the function to make the function non-negative.

We define the following.

Definition 1.31. The *capacity* $c(\wp)$ of the transmission is

$$c(\wp) = 1 + \wp \, \log_2 \, \wp + (1 - \wp) \, \log_2 (1 - \wp),$$

where $\wp < 1/2$. $\qquad \blacksquare$

The capacity is 1 if $\wp = 0$ and close to 0 if $\wp$ is close to $1/2$. We have the following proposition.

Proposition 1.32. *The capacity is a monotonic decreasing function of $\wp$ from 0 to $1/2$.*

Proof. Let $c(x) = 1 + x \log_2 x + (1 - x) \log_2(1 - x)$ be the capacity function. Then, we have

$$c'(x) = \log_2 x - \log_2(1 - x) = \log_2 \frac{x}{1 - x},$$

which is negative as long as $0 < x < 1/2$, hence $c(x)$ is monotonically decreasing between 0 and $1/2$. $\qquad\square$

Observe that the above definitions are consistent with all of the previous definitions given in the context of linear codes. Now, we may state *Shannon's theorem* [33].

Theorem 1.33 (Shannon's theorem). *For any rate of information $0 < R < $ capacity, there is a sequence of codes $\{M_n\}$ such that the rate of information of M_n, $R(M_n) > R$, and the average probability of incorrect decoding $P(M_n) \to 0$ as $n \to \infty$.* $\qquad\blacksquare$

We will skip the proof of the above existence theorem. Although it is interesting conceptually, there is no constructive proof of Shannon's theorem. In light of the above Shannon's theorem, observe that the $[2^n - 1, 2^n - n - 1, 3]$ Hamming codes $\mathbf{C_n}$ will have

$$R(\mathbf{C_n}) \to 1,$$

$$\delta = 3/(2^n - 1) \to 0,$$

i.e., the correcting power tends to zero as $n \to \infty$. Note that the above theorem states that there is a sequence $\{M_n\}$ with the average probability of incorrect decoding $P(M_n) \to 0$. The above sequence of $\delta \to 0$ of *Hamming codes* just show that their probability of correct decoding $1 - P(M_n) \to 0$ or of incorrect decoding $P(M_n) \to 1$. Or we know *Hamming code* can correct only one error; as the code gets longer and longer, the probability to have more than one error tends to be larger and larger, and then *Hamming code* would fail as the length of code becomes longer $P(M_n) \to 1$. In other words, Hamming codes are not the best code that we should expect. We may wish to construct other sequence of linear codes that do fulfill some of the expectations. It gives a powerful incentive to search for long and useful codes, and it separates the study of codes into two branches: theoretical codes disregarding the decoding procedures (we may use the Remark after Proposition 1.25 using brute force to search for its code word or we may use the universal maximum likelihood decoding) and the decoding procedures themselves. Certainly, for practical applications, we cannot allow $n \to \infty$. The interested reader of Shannon's theorem is referred to the work of van Lint [9].

1.9. Gilbert–Varshamov's Bound

Note that *Shannon's theorem* is an *existence* theorem. In this section, we construct a weaker theorem, the theorem of *Gilbert–Varshamov's Bound*. Let us consider the n-dimensional vector space, word space, $\mathbf{U} = \mathbf{F_q^n}$ over a finite field $\mathbf{F_q}$ of q elements. Furthermore, let vector $\alpha \in \mathbf{U}$ and $\mathbf{B}(\alpha, \mathrm{s})$ be the ball of radius $s \leq n$ around α, i.e.,

$$\mathbf{B}(\alpha, s) = \{\beta : d(\alpha, \beta) \leq s\}.$$

Let us define its volume $\mathbf{U}(\alpha, \mathrm{s})$ to be the number of elements inside $\mathbf{B}(\alpha, s)$. Note that if $\lfloor s \rfloor = r$, then $\mathbf{B}(\alpha, s) = \mathbf{B}(\alpha, r)$. We have the following geometric proposition.

Proposition 1.34. *Let r be an positive integer. We have $\mathbf{U}(\alpha, r) = \sum_{i=0}^{r} \mathbf{C}_i^n (q - 1)^i$. Note that it is independent of α, and we will denote the common number by $\mathbf{U}_q(n, r)$.*

Proof. Let $0 < d(\alpha, \beta) = i \leq r$. Then, α, β differ at i positions. There are $\mathbf{C}_i^n$ selections of the positions, and there are $(q - 1)^i$ possible values for those different positions and hence the proposition. $\square$

Let us consider a code $\mathbf{M} \subset F_q^n$ which may not be linear, and let m be the number of elements in $\mathbf{M}$. We will say that this is an (n, m, s) code, where n is the dimension of the word space and s is greater or equal to the minimal Hamming distance between d elements of $\mathbf{M}$. For a fixed pair (n, s) with $n \geq s \geq 0$, we may consider the maximal possible m as follows.

Definition 1.35. Let us use the notations of the preceding paragraph. Let $A(n, s) = \max\{m \mid \text{an } (n, m, d) \text{ code exists with } d \leq s\}$. ∎

Note that if $\lfloor s \rfloor = r$, then $A(n, s) = A(n, r)$. For a sequence of (n_i, m_i, d_i) codes, we may assume that they have the same rate of distance δ, i.e., $d_i = n_i \delta$. Then, we may consider **the limit of the rate of information of the sequence** to be $\lim_{i \to \infty} n_i^{-1} \log_q m_i$. More generally, we may define the following.

Definition 1.36. Let the **limit sup of the rate of information** $\alpha(\delta)$ be

$$\alpha(\delta) = \lim \ \sup \ n^{-1} \log_q A(n, \delta n).$$ ∎

We define the **entropy function** $H_q(x)$ as follows.

Definition 1.37. For $0 \le x \le \frac{q-1}{q}$, we define the *entropy function* $H_q(x)$ as

$$H_q(0) = 0,$$

$$H_q(x) = x \log_q(q-1) - x \log_q x - (1-x) \log_q(1-x), \qquad x \ne 0.$$

■

We need the following well-known Stirling's formula:

$$\log n! \mapsto n \log n - n + O(\log n) \text{ as } n \mapsto \infty.$$

We have the following propositions.

Lemma 1.38. *Let* $0 \le \lambda \le \frac{q-1}{q}$ *and* $q \ge 2$. *Then,*

$$\lim_{n \mapsto \infty} n^{-1} \log_q \mathbf{U}_q(n, \lfloor \lambda n \rfloor) = H_q(\lambda).$$

Proof. Let $r = \lfloor \lambda n \rfloor$. We separate the proof into two cases: Case 1. $q = 2$; Case 2. $q \ge 3$.

Case 1. We suppose that $q = 2$. Then, we have $\lambda \le \frac{1}{2}$, and

$$H_2(0) = 0,$$

$$H_2(x) = -x \log_2 x - (1-x) \log_2(1-x), \qquad x \ne 0$$

and

$$\mathbf{U}_2(n, r) = \sum_{i=0}^{r} \mathbf{C}_i^n.$$

Using Stirling's formula, we deduce that

$$n^{-1} \log_2 \left(\sum_{i=0}^{r} \mathbf{C}_i^n \right) \ge n^{-1} \log_2 \mathbf{C}_r^n = n^{-1} \log_2 \frac{n!}{r!(n-r)!}$$

$$= n^{-1}(n \log_2 n - r \log_2 r - (n-r) \log_2(n-r) + O(\log_2(n)))$$

$$= \log_2 n - (r/n) \log_2((r/n)n) - (1 - (r/n))$$

$$\times \log_2 ((1 - (r/n))n) + n^{-1} O(\log_2(n)).$$

Since $\lambda n - \epsilon = r$ for some $0 \le \epsilon < 1$, it is not hard to deduce that

$$n^{-1} \log_2 \mathbf{C}_r^n$$

$$\ge \log_2 n - \lambda \log_2(\lambda n) - (1 - \lambda) \log_2((1-\lambda)n) + n^{-1} O(\log_2(n))$$

$$\ge -\lambda \log_2(\lambda) - (1 - \lambda) \log_2((1 - \lambda)) + n^{-1} O(\log_2(n))$$

$$\ge H_2(\lambda) + n^{-1} O(\log_2(n)).$$

So, we prove one direction. For the other direction, we have

$$2^{-nH(\lambda)} = (\lambda)^{n\lambda}(1-\lambda)^{(1-\lambda)n}$$

$$= (1-\lambda)^n \left(\frac{\lambda}{1-\lambda}\right)^{\lambda n}.$$

Therefore, we have

$$2^{-nH(\lambda)} \sum_{0 \le i \le \lambda n} \mathbf{C}_i^n = \sum_{0 \le i \le \lambda n} \mathbf{C}_i^n (1-\lambda)^n \left(\frac{\lambda}{1-\lambda}\right)^{\lambda n}$$

$$\le \sum_{0 \le i \le \lambda n} \mathbf{C}_i^n (1-\lambda)^n \left(\frac{\lambda}{1-\lambda}\right)^i$$

$$= \sum_{0 \le i \le \lambda n} \mathbf{C}_i^n \lambda^i (1-\lambda)^{n-i}$$

$$\le (\lambda + (1-\lambda))^n = 1$$

or

$$\sum_{0 \le i \le \lambda n} \mathbf{C}_i^n \le 2^{nH(\lambda)}.$$

After we take $\log_2$ on the above inequality, our Case 1 follows.

Case 2. We suppose that $q \ge 3$ or $(q-1) \ge 2$. Then, for $i \le r$, we always have

$$\mathbf{C}_{i-1}^n (q-1)^{i-1} < \mathbf{C}_i^n (q-1)^i.$$

For instance, for a proof, we have

$$\mathbf{C}_{i-1}^n (q-1)^{i-1} < \mathbf{C}_i^n (q-1)^i \Leftrightarrow i < (q-1)(n-i+1)$$

$$\Leftrightarrow qi < (q-1)(n+1) \Leftrightarrow i < (n+1)\frac{q-1}{q}$$

$$\Leftarrow i < n\frac{q-1}{q} \Leftarrow i \le r.$$

It follows from Proposition 1.34 and the above that

$$\mathbf{C}_r^n (q-1)^r \le \mathbf{U}_q(n,r) \le (1+r)\mathbf{C}_r^n (q-1)^r.$$

Now, we simply imitate the proof of Case 1 to deduce Case 2 (see Exercises). $\qquad\square$

Lemma 1.39. *Let n, d be positive integers and $d < n$. Then, we have*

$$A(n,d) \ge q^n / U_q(n,d-1) \ge q^n / U_q(n,d).$$

Proof. The last inequality is obvious, since $U_q(n, d) \geq U_q(n, d - 1) > 0$. Let **C** be a code whose number m of elements gives $A(n, d)$. Then, we know that there is no word in F_q^n with distance d or more to all words in **C**. Otherwise, we may throw in the new element and increase the number m by 1. This is contradictory to the definition of m. Therefore, we have the following:

$$A(n, d) \times U_q(n, d - 1) \geq q^n. \qquad \square$$

It follows from Lemma 1.39 that $A(n, s) = A(n, \lfloor \lambda s \rfloor) \geq q^n / U_q(n, \lfloor \lambda s \rfloor)$.

Proposition 1.40 (Gilbert–Varshamov's bound). *We have*

$$\alpha(\delta) \geq 1 - H_q(\delta).$$

Proof. We have

$$\alpha(\delta) = \lim \sup \ n^{-1} \log_q A(n, \delta n)$$
$$= \lim \sup \ n^{-1} \log_q A(n, \lfloor \delta n \rfloor)$$
$$\geq \lim_{n \to \infty} (1 - n^{-1} \log_q \ U_q(n, \lfloor \delta n \rfloor))$$
$$= 1 - H_q(\delta). \qquad \square$$

We may compare Proposition 1.40 with Shannon's theorem. Consider the binary case, i.e., $q = 2$, then an $(n, m, n\delta)$ code with minimal Hamming distance $n\delta$ may correct $\frac{n\delta - 1}{2}$ errors (cf. Proposition 1.23). In this case, we observe that the rate of correcting error $\frac{n\delta - 1}{2n} = \frac{\delta}{2} - \frac{1}{2n}$ for large values of n. To compensate for a noisy channel of having a **rate of errors** $\wp$, we should choose a code with $\delta = 2\wp$. Then, we have the following equations (note that $\log_2(2 - 1) = 0$ and recall that $\alpha(\delta) = \alpha(2\wp)$ is the limit sup of the rate of information):

$$\alpha(2\wp) \geq 1 + 2\wp \log_2 \ 2\wp + (1 - 2\wp) \ \log_2(1 - 2\wp).$$

Gilbert–Varshamov's bound

For comparison, in Shannon's theorem, we have the rate of information $R(M)$:

$$R(M) \geq 1 + \wp \ \log_2 \ \wp + (1 - \wp) \ \log_2(1 - \wp) - \epsilon.$$

Shannon's theorem

Recall that $1 + x \log_2 x + (1 - x) \log_2(1 - x)$ is the capacity function (Definition 1.31), which is a monotonic decreasing function from 0 to $1/2$

(Proposition 1.32); we may deduce that the right-hand side number (drop the small ϵ) in the second equation using Shannon's theorem is always bigger than the corresponding number in the first equation as long as $2\wp < \frac{1}{2}$, which is valid for interested symmetric channels (see Exercise (5)). Therefore, we conclude that the rate of information $R(M)$ by Shannon's theorem satisfies a stronger inequality than the limit sup of the rate of information $\alpha(2\wp)$ by Gilbert–Varshamov's bound. Hence, Shannon's theorem is stronger in this case. However, Shannon's theorem is an existence theorem, and Gilbert–Varshamov's bound is constructive. Therefore, each has its own advantage. For more than 20 years, the Gilbert–Varshamov's bound, which serves as a standard to measure all codes, has been met by the *classical Goppa codes* (cf. Proposition 3.22), and it has only been surpassed by the *geometric Goppa codes* in the 1990s (see Theorem 5.11, Part IV).

Exercises

(1) Let $\mathbf{C_n}$ be the Hamming codes $[2^n - 1, 2^n - n - 1, 3]$. Let R_n be their rates of information. Find $\lim_{n \mapsto \infty} R_n$.

(2) Let $\mathbf{C_n}$ be the triple repetition code of length $3n$, i.e., a message $[a_1 a_2 \cdots a_n]$ is sent to $[a_1 a_1 a_1 a_2 a_2 a_2 \cdots a_n a_n a_n]$. Let R_n be their rates of information. Find $\lim_{n \mapsto \infty} R_n$.

(3) Find the maximal value of $H_q(x) - x$.

(4) Finish the proof of Case 2 in Lemma 1.37.

(5) Prove the last paragraph of Section 1.9.

PART II

Ring Codes

Chapter 2

Rings

If we treat letters as unrelated symbols, then we have information theory which cannot correct any error. If we treat sequences of letters as vectors in a vector space, then we have Hamming codes which can correct one error. Naturally, we are interesting in codes which may correct multiple errors. Therefore, we are very likely to involve more algebraic relations. Furthermore, it follows from *Shannon's Theorem* that we may have to work with long codes to make better codes. However, a long code will introduce complexity for decoding. To overcome this complexity, we have to assume more algebraic structures to help us.

To correct more errors, and to work on longer codes, we have to study the polynomial rings $\mathbf{F}_q[\beta]$ over finite fields $\mathbf{F}_q$ and related topics. In this chapter, we cover the essential materials from ring theory and finite-field theory and related topics for future use.

2.1. Preliminaries

In mathematics, we have **Ring theory**, in addition to **Group theory**. By definition, a **ring R** is a mathematical object which satisfies the following definition.

Definition 2.1. A ring $(\mathbf{R}, +, \times)$ is a set $\mathbf{R}$ with two binary operations $+$ and $\times$, such that $\mathbf{R}$ is closed under those two operations, and they satisfy the following:

(1) *Group law:* $(\mathbf{R}, +)$ is an abelian group.
(2) *Associative law:* We always have $a \times (b \times c) = (a \times b) \times c$.

(3) *Distributive law:* We always have $a \times (b + c) = a \times b + a \times c$ and $(a + b) \times c = a \times c + b \times c$.

Sometimes, we use $\cdot$ in place of $\times$. If $a \cdot b = b \cdot a$ always, then we say **R** is a *commutative ring*. The simplest ring is a ring that consists of only 0. It is called the zero ring and is not of interest. In this book, we assume that any ring used is not a zero ring. If **R** is commutative, and there is a multiplicative identity $e \neq 0$ (we usually write e as 1), then we say **R** is a commutative ring with identity. In this book, all rings are **commutative rings with identity**, if not stated explicitly otherwise. ■

Remark: (1) In a ring **R**, if $e = 0$, then $a = a \cdot e = a \cdot 0 = 0$ for any element a, and the ring is a zero ring.

(2) A non-zero element α in **R** which is not a zero divisor, i.e., $\beta\alpha = 0$ or $\alpha\beta = 0 \Rightarrow \beta = 0$, will be called a *regular element.*

(3) It is easy to see that a commutative ring with identity **R** which contains a field **K** must be a vector space over **K**. In some books, it is called **K**-ring. We shall use this terminology. In coding theory, we only involve **K**-rings. In this book, we assume that all rings are **K**-rings for some field **K**. Every **K**-ring **R** can be written as $\mathbf{K}[\{\mathbf{R}\}]$, i.e., a polynomial ring on the elements of **R**.

(4) The rings **R** used in coding theory usually contain some finite field $\mathbf{F}_q$ as the polynomial rings $\mathbf{F}_q[\{x\}]$, meromorphic power series rings $\mathbf{F}_q((\{x\}))$ or rational function rings $\mathbf{F}_q(\{x\})$ (see the following). Therefore, the rings **R** can be considered as a vector space over $\mathbf{F}_q$ with a multiplication between elements in $\mathbf{F}_q$ and **R**. ■

As usual, we define ideal **I** as a subring of the ring **R** with the property that $r \cdot \mathbf{I} = \{r \cdot a : \forall a \in \mathbf{I}\} \subset \mathbf{I}$. There is a canonical map $\pi : \mathbf{R} \mapsto \mathbf{R}/\mathbf{I} = \{r + \mathbf{I} : r \in \mathbf{R}\}$.

2.2. The Finite Field $\mathbf{F}_q$

Recall that *letters* in coding theory are picked up from a finite field. We wish to discuss some basic structure of a finite field **F**. One of the basic properties we assume is the following theorem from *Theory of Groups*, Vol. 1, p. 147, by Kurosh [12], where we consider abelian *additive* group; note that all theorems there will be applied in this book *multiplicatively.*

Theorem 2.2 (Fundamental Theorem of Finitely Generated Abelian Group). *Given any finitely generated abelian group G, then G is isomorphic to $(\oplus_i \mathbb{Z})(\oplus_i \mathbb{Z}_{c_i})$ as abelian additive group, where the $\oplus$ has only finitely many arguments and $c_i > 0$ for all i.*

Proof. See the reference. $\qquad\square$

We have the following easy propositions for abelian groups.

Proposition 2.3. (1) *Let $G = \mathbb{Z}_a \oplus \mathbb{Z}_b$ and $d = lcm(a, b)$. Then, every element $x \in G$ satisfies*

$$dx = 0.$$

(2) *Let $G = \mathbb{Z}_a \oplus \mathbb{Z}_b$, where a, b are co-prime. Then, G is cyclic of order ab.*

Proof. (1) It is easily provable. (2) Let c, d be selected such that $ac + bd = 1$. Let x be a generator of $\mathbb{Z}_a$ and y be a generator of $\mathbb{Z}_b$. We claim that $z = (dx, cy)$ is a generator $\mathbb{Z}_a \oplus \mathbb{Z}_b$.

We have the following computations:

$$az = (adx, acy) = (0, acy) = (0, (1 - bd)y) = (0, y + (-bd)y) = (0, y).$$

Similarly, we have $bz = (x, 0)$. Therefore, z generates G, and G is cyclic. It is easy to show that $order(G) = ab$. $\qquad\square$

Note that if we write the group operation multiplicatively, the equation $dx = 0$ in (1) will be written as $x^d = 1$.

Proposition 2.4. *Let G be a multiplicative subgroup of a finite field $\mathbf{F}_q$. Then, G is cyclic.*

Proof. Since $\mathbf{F}_q$ is finite, G is a finitely generated abelian group. Let us make another copy of G with the product replaced by summation. We may simply call it G^0. By Theorem 2.2, we have G^0 to be isomorphic to $\oplus(\mathbb{Z}_{c_i})$. Note that this isomorphism will change the multiplication of G to the summation in $\oplus(\mathbb{Z}_{c_i})$. Let us make an induction on the number factors n. In general, let us consider any finite group G, which can be written as $\oplus_{i=1}^n(\mathbb{Z}_{c_i})$. If $n = 1$, then our proposition is clearly true since 1 will be a generator of $\mathbb{Z}_{c_i}$, and $\mathbb{Z}_{c_i}$ is cyclic. Let us assume that the proposition is true for any abelian subgroup with number $m < n$ of factors. If there are at least two a_i, a_j with $d = lcm(a_i, a_j) < a_i a_j$, then all elements in the subgroup $(\mathbb{Z}_{a_i} \oplus \mathbb{Z}_{a_j})$ which has $a_i a_j$ elements satisfy

$$dx = 0,$$

which again translates to the multiplication of the field:

$$x^d = 1.$$

Then, the above polynomial will have more than d distinct elements as solutions, which is impossible for a field. Now, it follows from (1) of Proposition 2.3 that $\mathbb{Z}_{c_i} \oplus \mathbb{Z}_{c_j} = \mathbb{Z}_{c_i c_j}$. After this recombination, the factorization has only $n - 1$ factors, and our proposition follows. $\qquad \square$

One interesting criterion for a polynomial $f(x) \in \mathbf{F}[x]$ to have a multiple root in an algebraic closure Ω of $\mathbf{F}$ is the derivative test. We have to define the *derivative* algebraically.

Definition 2.5. Let $f(x) = \sum_i a_i x^i$. Then, the derivative $f'(x)$ is defined to be $f'(x) = \sum_i i a_i x^{i-1}$. $\qquad \blacksquare$

It is easy to see that the derivative obeys the following formal rules.

Proposition 2.6. *Let $a \in \mathbf{F}$ be a constant, and $f(x), g(x) \in \mathbf{F}[x]$. The derivative operation has the following rules:*

$$(1) : a' = 0,$$

$$(2) : (f(x) + g(x))' = f'(x) + g'(x),$$

$$(3) : (f(x)g(x))' = f'(x)g(x) + f(x)g'(x).$$

Proof. The proof is a routine check. $\qquad \square$

We have the following criterion.

Proposition 2.7. *An element $\beta \in \Omega$ is a multiple root of $f(x) \in \mathbf{F}[x] \iff \beta$ is a common root of $f(x)$ and $f'(x)$.*

Proof. Let β be a multiple root of $f(x)$. Then, $f(x) = (x - \beta)^2 h(x)$ with $h(x) \in \Omega[x]$. Then, we have

$$f'(x) = 2(x - \beta)h(x) + (x - \beta)^2 h'(x) = (x - \beta)g(x).$$

It is easy to prove that $g(x) \in \mathbf{F}[x]$. Therefore, β is a root of $f(x)$ and $f'(x)$. On the other hand, if β is a root of $f(x)$ and $f'(x)$, we write

$f(x) = (x - \beta) r(x)$. We have

$$f'(x) = r(x) + (x - \beta) r'(x).$$

We conclude that β must be a root of $r(x)$, and $f(x) = (x - \beta)^2 s(x)$. $\quad\square$

We have the following proposition which gives us a lot of information about the structures of finite fields.

Proposition 2.8. *Let* $\mathbf{F}_q$ *be a finite field of* $q = p^n$ *elements in an algebraic closure* Ω *of* $\mathbb{Z}_p$. *Then,* $\mathbf{F}_q^* = \mathbf{F}_q \setminus 0$ *is a multiplicative cyclic group of order* $p^n - 1$, *and* $\mathbf{F}_q$ *is the solution set in* Ω *of the following equation:*

$$x(x^{p^n - 1} - 1) = x^{p^n} - x = 0.$$

Furthermore, all solutions of the following equation in Ω *are distinct and a collection of the solutions is a field:*

$$x^{p^n} - x = 0.$$

If $\mathbf{L}$ *is another finite field in* Ω *with* p^m *elements with* $n|m$, *then* $\mathbf{L} \supset \mathbf{F}_q$, *and*

$$x^{p^n - 1} - 1 | x^{p^m - 1} - 1.$$

Proof. The first statement follows trivially from the preceding proposition. For the second statement, let $\mathbf{K}_1$ be equal to the collection of all solutions of the equation $f(x) = x^{p^n} - x = 0$ in Ω. Since Ω is an algebraic closure of $\mathbb{Z}_p$, then the equation $f(x) = 0$ splits completely. By considering the derivative of $f'(x) = 1$, the equation has no multiple root. Therefore, $\mathbf{K}_1$ consists of p^n elements. Moreover, let $y, z \in \mathbf{K}_1$, then we have

$$(y + z)^{p^n} = y^{p^n} + z^{p^n} = y + z, \quad \text{and} \quad (yz)^{p^n} = y^{p^n} z^{p^n} = yz.$$

Therefore, $y + z, yz \in K$. It is easy to check that all requirements of a field are satisfied to establish that $\mathbf{K}_1$ is a field of p^n elements and thus equal to $\mathbf{F}_q$. For the last part of the proposition, let $m = ns$, then we have $p^m - 1 = p^{sn} - 1 = (p^n - 1)r$. Therefore, $x^{p^m - 1} - 1 = x^{(p^n - 1)r} - 1 = (x^{p^n} - 1)g(x)$. Thus, $\mathbf{L} \supset \mathbf{F}_q$. $\quad\square$

It is not hard to see the following proposition.

Proposition 2.9. *We have* $\Omega = \cup_{m>0} \mathbf{F}_{p^m}$.

Proof. Let $\beta \in \Omega$. Then, $\mathbb{Z}_p[\beta]$ is a finite extension of $\mathbb{Z}_p$. Therefore, $\mathbb{Z}_p[\beta] = \mathbf{F}_{p^m}$ for some m. $\quad\square$

Let us have the following common definition.

Definition 2.10. Let a mapping ρ be defined as $\rho(\alpha) = \alpha^p$. The mapping ρ is called the *Frobenius map*. ∎

We have the following usual proposition.

Proposition 2.11. *The Frobenius map ρ is an automorphism of $\mathbf{F}_q$ of order m, where $q = p^m$.*

Proof. It is a routine check to see that ρ is an automorphism since all elements β are solutions of the equation $x^{p^m} - x = 0$. Then, all elements $\beta \in \mathbf{F}_q$ satisfy

$$\beta^{p^m} = \beta$$

and $\rho^m(\beta) = \beta^{p^m} = \beta$. Therefore, $\rho^m = id$. On the other hand, if $\rho^\ell = id$ for some $\ell < m$, then we have $\beta^{p^\ell} = \beta$ for all $\beta \in \mathbf{F}_q$, which is impossible. □

Corollary 2.12. *Let $m = \ell s$. Then, ρ^ℓ is an isomorphism of $\mathbf{F}_{p^m}$ which fixes $\mathbf{F}_{p^\ell}$. Note that the map ρ^ℓ as an isomorphism of $\mathbf{F}_{p^m}$ is of order s. Furthermore, it follows from Galois theory that any automorphism of $\mathbf{F}_{p^m}$ which fixes $\mathbf{F}_{p^\ell}$ is of the form $(\rho^\ell)^j$ for some suitable j.* ∎

We *assume Galois theory* and the above corollary. We define the following.

Definition 2.13. A field $\mathbf{K}$ is said to be *perfect* if either its characteristic is 0 or every element in $\mathbf{K}$ has a pth root in $\mathbf{K}$, where p is its characteristic. ∎

Proposition 2.14. *A finite field is perfect.*

Proof. It is easy to see that the Frobenius map ρ is one-to-one. The pigeon hole principle implies ρ is onto. Therefore, for any given $\alpha \in \mathbf{K}$, there is an $\beta \in \mathbf{K}$ such that

$$\beta^p = \alpha.$$

Clearly, β is the pth root of α. □

2.3. The Computer Programs for Finite Fields

The coding and the decoding processes depend on the computer programs heavily. This section is written for those readers who are not familiar with computer programs, especially a student of pure mathematics.

Let us assume that $p = 2$ in this section. The reader is requested to discuss the cases of $p > 2$. Let the set $\mathbf{U}_m$ of all m-bits be a vector space of dimension m over $\mathbf{F}_2$. We want to provide a more algebraic structure to $\mathbf{U}_m$. In fact, a finite field $\mathbf{F}_{2^m}$ over $\mathbf{F}_2$ can be used to represent $\mathbf{U}_m$. This is the right way to generalize a vector space in coding theory. The addition is simple, while the multiplication is very messy in practice. We shall manage the multiplication in three ways as follows.

(1) **The table of logarithm:** We shall only consider examples. The reader may generalize the following construction process to the general setup for any prime field $\mathbb{Z}/p\mathbb{Z}$ with $p > 2$ and any finite field $\mathbb{Z}_{p^m}$.

Let us consider $\mathbf{F}_{2^4}$ over F_2. Let us write $F_{2^4} = F_2[\alpha]$, where α satisfies the equation $x^4 + x + 1 = 0$. We have the following list of powers of α: $\{1 = \alpha^0, \alpha, \alpha^2, \alpha^3, \alpha + 1 = \alpha^4, \alpha^2 + \alpha = \alpha^5, \alpha^3 + \alpha^2 = \alpha^6, \alpha^3 + \alpha + 1 = \alpha^7, \alpha^2 + 1 = \alpha^8, \alpha^3 + \alpha = \alpha^9, \alpha^2 + \alpha + 1 = \alpha^{10}, \alpha^3 + \alpha^2 + \alpha = \alpha^{11}, \alpha^3 + \alpha^2 + \alpha + 1 = \alpha^{12}, \alpha^3 + \alpha^2 + 1 = \alpha^{13}, \alpha^3 + 1 = \alpha^{14}, 1 = \alpha^{15}\}$.

It is clear that $1, \alpha, \alpha^2, \alpha^3$ are linearly independent over F_2. Let us write them as $[1000], [0100], [0010], [0001]$, then the above list of powers of α may be re-written as $[1000], [0100], [0010], [0001], [1100], [0110], [0011], [1101], [1010], [0101], [1110], [0111], [1111], [1011], [1001], [1000]$. One way to treat the problem of multiplication is to look up the list and find i, j for any two elements $[a_1a_2a_3a_4]$, $[b_1b_2b_3b_4]$ such that

$$[a_1a_2a_3a_4] = \alpha^i,$$
$$[b_1b_2b_3b_4] = \alpha^j,$$

and we define

$$\log_\alpha([a_1a_2a_3a_4]) = i,$$
$$\log_\alpha([b_1b_2b_3b_4]) = j.$$

Then, we have

$$\log_\alpha([a_1a_2a_3a_4][b_1b_2b_3b_4]) = i + j.$$

Let us find the residue value k of $i + j$ module $15 = 2^4 - 1$, i.e., $k = i + j \mod 15$ and $0 \leq k < 15$. Now, we may look up the above list of

powers again to find the corresponding value of α^k to determine the value
of $[a_1a_2a_3a_4][b_1b_2b_3b_4]$. This method consists of looking up **the table of
logarithm** of length $2^4 - 1$ (in general, for the field F_{2^m} is a table of length
$2^m - 1$. If m is large, say $m > 32$, then the table will be beyond the memory
of an ordinary computer and not be feasible).

(2) **The table of multiplication:** Let us consider the implementation
of a finite field **K** of p^m elements. We may consider the pair $\mathbf{K} \supset \mathbb{Z}_p$. We
shall implement the summation and the multiplication for the prime field
$\mathbb{Z}_p$ first. The summation shall be the usual sum for integers and then mod
p. For multiplication, there are two ways, either we multiply two numbers
and then mod p, or we use a multiplication table, if the number p of the
elements in $\mathbb{Z}_p$ is not too big, and then we look up the table to achieve the
multiplication. Note that the size of the table is $(1/2)(p-1) \times (p-2)$ after
deleting all multiplications of the form $0 \times a = a \times 0 = 0$ and $1 \times a = a \times 1 = a$,
and then, the size is the determining factor for the decision.

Let us program the field **K** which is uniquely determined by the number
p^m of elements in **K**, i.e., let **K** be the collection of all vectors of the form
$[a_1, a_2, \ldots, a_m]$ where $a_j \in \mathbb{Z}_p$. We define the summation as the summation
of vectors. For multiplication, consider the polynomial $x^{p^m} - x$. There are
many monic polynomials factor $f(x) \in \mathbb{Z}_p[x]$ which are irreducible over $\mathbb{Z}_p$
and of degree m (see Exercises). Let us select such a polynomial $f(x)$. Let
$v = [a_0, a_1, \ldots, a_{m-1}]$ and $u = [b_0, b_1, \ldots, b_{m-1}]$ be two element in **K**, then
we write

$$f_v = a_0 + a_1 x + a_2 x^2 + \cdots + a_{m-1} x^{m-1},$$
$$f_u = b_0 + b_1 x + b_2 x^2 + \cdots + b_{m-1} x^{m-1}$$

and compute

$$f_v \times f_u \quad \mod (f(x)) = f_{vu}(x) = \sum_{j=0}^{m-1} c_j x^j,$$

and we *define*: $v \times u = [c_0, c_1, \ldots, c_{m-1}]$. We have two ways to define the
multiplication for the field **K**: If $(1/2)(p^m - 1)(p^m - 2)$ is small enough, then
we may form a table of multiplication. Or if $(1/2)(p^n - 1)(p^n - 2)$ is too
big, then for every multiplication, we shall go through the above equation
of modulo polynomial $f(x)$ to find the product (see (3)).

Let us discuss the following example.

Example 1: Let **K** be a finite field of 2^3 elements. Then, $\mathbf{K} \supset \mathbb{Z}_2$, and
K is a vector space of dimension 3 over $\mathbb{Z}_2$. Any element in **K** can be

represented as $[a_0, a_1, a_2]$, where $a_j = 0, 1$. Let consider the polynomial $f(x) = 1 + x + x^3$. It has no root in $\mathbb{Z}_2$; therefore, it is irreducible over $\mathbb{Z}_2$ (see Exercises). Henceforth, we may use it to define the multiplication in **K**. For the sake of notations, let us use the ordinary numerals $\{0, 1, 2, \ldots, 7\}$, with each written in binary expansion, and use zeroes to fill each to the length 3 as follows:

$$0 = [0,0,0], \ 1 = [1,0,0], \ 2 = [0,1,0], \ 3 = [1,1,0],$$

$$4 = [0,0,1], \ 5 = [1,0,1], \ 6 = [0,1,1], \ 7 = [1,1,1].$$

On the other hand, we represent $[a_0, a_1, a_2]$ as $a_0 + a_1 x + a_2 x^2$, i.e., we represent $[0,1,0]$ as x and $[0,0,1]$ as x^2. Then, we have $2 \times 2 \ (= x \cdot x = x^2) = 4$. Similarly, we may compute the following table.

$\times$	2	3	4	5	6	7
2	4	6	3	1	7	5
3	6	5	7	4	1	2
4	3	7	6	2	5	1
5	1	4	2	7	3	6
6	7	1	5	3	2	4
7	5	2	1	6	4	3

For any other finite field, we use the above method to construct a multiplication matrix $(a_i a_j)$ for all pairs of non-zero elements (a_i, a_j) in the field. Once we have the table, we have to lookup the table once to find out the result of multiplication of the pair (a_i, a_j).

Note that the table is symmetric, and the multiplication depends on the polynomial $f(x) = x^3 + x + 1$, for instance, $5 \times 3 = 4$ which comes from the following equation:

$$(x^2 + 1)(x + 1) = x^3 + x^2 + x + 1 = x^3 + x + 1 + x^2 = x^2 \quad \text{mod } (x^3 + x + 1).$$

Note that $x^{8-1} - 1 = (x + 1)(x^3 + x + 1)(x^3 + x^2 + 1) \mod 2$. The other possible selection of the irreducible polynomial of degree 3 is $x^3 + x^2 + 1$. If we use it to define the multiplication, then we have $5 \times 3 = 2$. It is easy to see that the definition of multiplication depends on the selection of the irreducible polynomial of degree 3. ∎

(3) **Linear Feedback Shift Register:** Let us consider another method. Let us assume that $p = 2$ in this subsection. Let us consider F_{2^4}. Let the

defining equation be

$$f(x) = x^4 + x + 1 = 0.$$

Note that we have $1 = -1$. Let $F_{2^4} = F_2[\alpha]$ and α satisfies the above equation. Let us represent $[a_1 a_2 a_3 a_4]$ as $a_1 + a_2\alpha + a_3\alpha^2 + a_4\alpha^3$ and $[b_1 b_2 b_3 b_4]$ as $b_1 + b_2\alpha + b_3\alpha^2 + b_4\alpha^3$. Then certainly, $[a_1 a_2 a_3 a_4][b_1 b_2 b_3 b_4]$ can be represented as $(a_1 + a_2\alpha + a_3\alpha^2 + a_4\alpha^3)(b_1 + b_2\alpha + b_3\alpha^2 + b_4\alpha^3) = (a_1 b_1) + (a_1 b_2 + a_2 b_1)\alpha + \cdots + (a_4 b_4)\alpha^6 = c_1 + c_2\alpha + \cdots + c_7\alpha^6$. Now, we want to re-write the last expression as a polynomial in α of degree at most 3 by going mod the defining equation. In general, we may reduce any polynomial $\sum_{j+1}^{\ell} c_i \alpha^{i+1}$ to a polynomial of degree at most 3 using a *linear feedback shift register*, which simulates the process of modulo out the defining equation as follows:

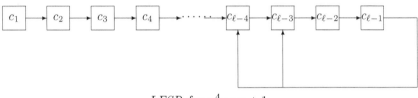

LFSR for $x^4 = x + 1$.

The process is as follows: Assume $c_\ell = 1 \neq 0$, then we push all terms one step rightward, and we make the change $c_{\ell-4} \to c_{\ell-4} + 1$ and $c_{\ell-3} \to c_{\ell-3} + 1$. The remaining terms stay the same. The next step is pushing all terms one step rightward. The term $c_{\ell-1}$ will fall off the horizontal line. If $c_{\ell-1}$ is 0, then we do nothing, and we push again. If $c_{\ell-1}$ is 1, then we use the above diagram to feed back and make changes of the terms. We keep making shifts until there are only four terms, c_1, c_2, c_3, c_4, left. What is left is the result of multiplication.

Exercises

(1) Let Ω be an algebraic closure of $\mathbf{F}_p$, where p is a prime number. Show that there is a unique field of order $p^m \subset \Omega$.

(2) Find a quadratic irreducible equation $f(x)$ over $\mathbf{F}_p$.

(3) Show that a polynomial $f(x)$ of degree n is irreducible over $\mathbf{F}_p \Leftrightarrow$

    (1) $f(x) \mid x^{p^n} - x$,

    (2) $(f(x), x^{p^m} + 1) = (1), \quad for \ all \ m < n$.

(4) Let α be a root of $x^2 + x + 1 = 0$ and β be a root of $x^2 + x + \alpha = 0$ in an algebraic closure Ω of $\mathbf{F}_2$. Show that $\mathbf{F}_4 = \mathbf{F}_2[\alpha]$ and $\mathbf{F}_{16} = \mathbf{F}_4[\beta]$.

(5) We use the notations of the preceding problem. Show that every element $\gamma \in \mathbf{F}_{16}$ can be expressed as $\gamma = a_3\alpha\beta + a_2\beta + a_1\alpha + a_0$, where $a_i = 0, 1$ are elements in $\mathbf{F}_2$. We may represent γ by an integer $a_3 2^3 + a_2 2^2 + a_1 2 + a_0$. Find γ^2 if γ is represented by 13.

(6) Find all group generators for $\mathbf{F}_{2^4}$.

(7) Over the finite field $\mathbf{F}_{p^n}$, show that

$$(\alpha + \beta)^p = \alpha^p + \beta^p.$$

(8) Find an element α which generates the multiplicative group $\mathbf{F}_{p^2}$, and find its defining equation.

(9) Let $\delta \in \mathbf{F}_{2^4}$ be a non-zero element. Find δ^{-1}.

2.4. The Total Quotient Rings

In the *classical Goppa code*[1] (see Section 3.4), we use the total quotient rings. Given a ring $\mathbf{R}$ without non-zero zero divisors, which is called an *integral domain*, we may consider the *quotient field* of $\mathbf{R}$ defined as the set s/r with $s \neq 0$ and $s/r \equiv s'/r'$ iff $sr' = s'r$. In the classical case of the integral domain $\mathbb{Z}$, the ring of integers, its quotient field is the the field of rational numbers $\mathbb{Q}$. A possible generalization of quotient field to the case of rings which are not integral domains are the total quotient rings.

Definition 2.15. If $\mathbf{R}$ is a ring with S equal to the set of regular elements, where S is a non-empty set, then a **total quotient ring of R** is a ring $\mathbf{F}$, which is $R \times S$ (where the element $[r, s]$ is written as r/s) with the equivalent relation that $\mathbf{r}/\mathbf{s} = \mathbf{r}'/\mathbf{s}'$ iff $rs' = s'r$. Its identity is the class of $\mathbf{b}/\mathbf{b}$ for all regular elements $\mathbf{b}$. The set S is called the set of *denominators*. Then, the quotient ring, $\{\frac{r}{s} : r \in R, s \in S\}$, is called the *total quotient ring* $\mathbf{S^{-1}R = RS^{-1}}$. ∎

The following examples are helpful.

Example 2: Let $\mathbb{R}$ be the ring of real numbers. Then, its total quotient ring is itself. ∎

---

[1] Valery Goppa (1939–), Soviet and Russian mathematician.

Example 3: Let **R** be the residue class ring $\mathbf{K}[x]/(g(x))$. Then, an element $\overline{f(x)} \in R$ is a regular element $\Leftrightarrow g(x)$ and $f(x)$ are co-prime. In that case, there are elements $h(x)$ and $r(x)$ such that

$$h(x)g(x) + f(x)r(x) = 1.$$

Therefore,

$$\frac{1}{\overline{f(x)}} = \overline{r(x)}.$$

It is easy to see that the total quotient ring of $\mathbf{K}[x]/(g(x))$ is itself. We apply the total quotient ring to the *classical Goppa code*. Especially, if $f(x) = \prod(x - \gamma_i)$ with $g(\gamma_i) \neq 0$, where $\{\gamma_i\}$ are all distinct, then any element of the form $\sum \frac{c_i}{(x-\gamma_i)} \in \mathbf{K}[x]/(g(x))$ can be written as $\overline{n(x)/f(x)}$ with $\deg(n(x)) < \deg(g(x))$. ∎

2.5. The Ring F[x]

As we point out that the rings used in coding theory are usually **K**-rings (see Chapter 3), it is easy to see that **R** can be expressed as $\mathbf{K}[\{\mathbf{R}\}]$. The simplest and most useful polynomial ring in coding theory is $\mathbf{K}[x]$, where x is a variable. Let us use **R** to denote the polynomial ring $\mathbf{F}[x]$ for any field **F**. We have the following definition.

Definition 2.16. Let us consider **R**. Let a polynomial $f(x) = \sum_i c_i x^i$ be given. The degree function, $\deg(f(x))$, of any polynomial $f(x)$ is as follows:

$$\deg(f(x)) = \begin{cases} \max\{i : c_i \neq 0\} & \text{if } f(x) \neq 0 \\ -\infty & \text{if } f(x) = 0. \end{cases}$$ ∎

Using the fact that the field **F** is an integral domain, we have the following basic properties of $\deg(f(x))$:

$$\deg(f(x)g(x)) = \deg(f(x)) + \deg(g(x)),$$

$$\deg(f(x) + g(x)) \leq \max(\deg(f(x)), \deg(g(x))),$$

and it shows that $\mathbf{F}[x]$ is an integral domain.

We have the following proposition.

Proposition 2.17. *Let $g(x)$, $\alpha_r(x)$, $\alpha_u(x)$, $\omega_r(x)$, $f_u(x)$ be polynomials in* **R**. *If maximal*$(\deg(\alpha_u(x)\omega_r(x)), \deg(\alpha_r(x)f_u(x))) < \deg(g(x))$, *then* $\alpha_u(x)\omega_r(x) \equiv \alpha_r(x)f_u(x) \mod (g(x)) \Rightarrow \alpha_u(x)\omega_r(x) = \alpha_r(x)f_u(x)$.

Proof. The proof is trivial. □

The following proposition is classically known, and the proof is left to the reader.

Proposition 2.18 (Euclidean algorithm). *Given any two polynomials $f_1(x)$, $f_2(x) \neq 0$ in* **R**. *There are polynomials $\beta_1(x)$ and $f_3(x)$ in* **R** *with $\deg(f_3(x)) < \deg(f_2(x))$ (note that $f_3(x)=0 \Leftrightarrow \deg(f_3(x)) = -\infty$) and $\deg(\beta_1(x)) \leq \deg(f_1(x))- \deg(f_2(x))\}$ such that*

$$f_1(x) = \beta_1(x)f_2(x) + f_3(x).$$ ■

The above proposition can be used to find the GCD of two polynomials $f(x)$, $g(x)$. The following process, known as the "long algorithm", is fundamentally important and is of interest for *decoding* purposes later in the book. We pay attention to the degrees of the polynomials involved, which turns out to be important in *decoding programs*. Note that *the Euclidean algorithm* is fast in computing. The main point in the process of decoding is that it will be modified to *an Euclidean algorithm with a stopping strategy* depending on the degrees involved (cf. Proposition 3.12).

Proposition 2.19 (Long algorithm). *Let $f_1(x), f_2(x), f_3(x)$ be given as in the preceding proposition. If $f_3(x) \neq 0$, the following sequence of polynomials $f_i(x)$ for $i = 3, \ldots, s$ with all polynomials non-zero, except possibly $f_s(x)$, can be defined inductively for $j = 3, \ldots, s - 1$ as*

$$f_{j-1} = \beta_{j-1}(x)f_j(x) + f_{j+1}(x)$$

such that after we name $n_j = \deg(f_j(x))$, $m_j = \deg(\beta_j(x))$, we have (1) $n_{j+1} < n_j$ and (2) $m_{j-1} = n_{j-1} - n_j$. Furthermore, after repeated back-substitution (see Proof), we have the following equation for $j \geq 3$:

$$\alpha_j(x)f_1(x) + \gamma_j(x)f_2(x) = f_j(x),$$

with $\deg(\alpha_j(x)) \leq n - n_{j-1}$ and $\deg(\gamma_j(x)) \leq n - n_{j-1}$, where $n = \max(\deg(f_1(x)), \deg(f_2(x)))$.

Proof. We may apply the Euclidean algorithm to the pair $f_2(x)$, $f_3(x)$ and so on, until we reach the case that $f_s(x) = 0$. The first part (1), (2) of the proposition is routine. Suppose that we have the second part for $j = 4, \ldots, \ell$ with $\ell < s$. Then, we have the following three equations:

$$f_{\ell-1} = \beta_{\ell-1}(x)f_\ell(x) + f_{\ell+1}(x),$$

$$\alpha_{\ell-1}(x)f_1(x) + \gamma_{\ell-1}(x)f_2(x) = f_{\ell-1}(x),$$

$$\alpha_\ell(x)f_1(x) + \gamma_\ell(x)f_2(x) = f_\ell(x).$$

Substituting the last two equations to the first one and collecting coefficients, we get the following equations:

$$\alpha_{\ell+1}(x) = \alpha_{\ell-1}(x) - \alpha_\ell(x)\beta_\ell(x),$$

$$\gamma_{\ell+1}(x) = \gamma_{\ell-1}(x) - \gamma_\ell(x)\beta_\ell(x).$$

Note that

$$\deg(\alpha_{\ell-1}(x)) \le n - n_{\ell-2} \le n - n_\ell,$$

$$\deg(\alpha_\ell(x)\beta_{\ell-1}(x)) \le n - n_{\ell-1} + n_{\ell-1} - n_\ell = n - n_\ell;$$

$$\deg(\gamma_{\ell-1}(x)) \le n - n_{\ell-2} \le n - n_\ell,$$

$$\deg(\gamma_\ell(x)\beta_\ell(x)) \le n - n_{\ell-1} + n_{\ell-1} - n_\ell = n - n_\ell.$$

Therefore, we have

$$\deg(\alpha_{\ell+1}(x)) \le n - n_\ell,$$

$$\deg(\gamma_{\ell+1}(x)) \le n - n_\ell. \qquad \square$$

A clever use of the above well-known proposition is a keystone of coding theory. Historically, the earliest application of the above theorem is in *number theory*. The continuous fraction of π is the following:

$$\pi = 3 + \cfrac{1}{7 + \cfrac{1}{15 + \cfrac{1}{1 + \cfrac{1}{292 + \cdots}}}}.$$

We get the sequence of numbers $\{3, \frac{22}{7}, \frac{333}{106}, \frac{355}{113}, \ldots\}$ by truncating the continuous fraction of π. As approximations of π, 3 was known to King Soloman, and found in a classical Chinese mathematics book, *Zhou Pai*, $\frac{22}{7} = 3 + \frac{1}{7}$ was known to Archimedes, and the number $\frac{355}{113} = 3 + \cfrac{1}{7 + \cfrac{1}{15 + 1}}$ was known as Zu's number (Zu, AD 340–501, China). From number theory, we know that this sequence gives us the best possible approximations of π.

For instance, if we restrict the denominator to be less than 1000, than $\frac{355}{113}$ is the best approximation to π.

When we apply the long algorithm to coding problems, we have to know the place to truncate the process. Now, Sugiyama, Kasahara, Hirasawa, and Namekawa [35] noticed the place to stop for the above *long algorithm* for a modern application for decoding purposes (see Proposition 3.12).

From the above proposition, we can deduce that $R = F[x]$ is a *principal ideal ring*, i.e., each ideal is generated by one element,

2.5.1. *LFSR*

Let us consider the problem of combining the above proposition with a computer. As well-known, the *Euclidean algorithm* can be implemented effectively using LFSR (linear feedback shift register) in any computer. Let us consider

$$f_1(x) = \beta_1(x) f_2(x) + f_3(x)$$

as in the statement of the *Euclidean Algorithm*. Let

$$f_1(x) = c_0 + c_1 x + c_2 x^2 + \cdots + c_{n_1-1} x^{n_1-1} + c_{n_1} x^{n_1},$$

$$f_2(x) = a_0 + a_1 x + \cdots + a_{n_2-1} x^{n_2-1} + x^{n_2}.$$

In general, polynomial divisions can be performed using LFSR. Let us consider a **simple case over $\mathbf{F}_2$**. We use the following LFSR to perform the division:

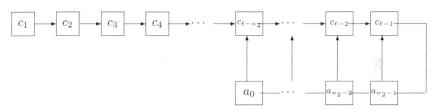

Let a polynomial $g(x) = c_0 + c_1 x + \cdots + c_\ell x^\ell$ be given, and we want to use $f_2(x)$ to cut down the degree ℓ of $g(x)$. If $\ell < n_2$, then we will not do anything. The process is to assume $\ell \geq n_2, c_\ell = 1 \neq 0$, then $c_{\ell-n_2} \to c_{\ell-n_2} + a_0$, $c_{\ell-n_2+1} \to c_{\ell-n_2+1} + a_1$, and so on. If $\ell - 1 \geq n_2$, the next step is pushing all terms one step rightward. The term $c_{\ell-1}$ will fall off the horizontal line. If it is 0, then we do nothing, and we push again. If $c_{\ell-1}$ is 1, then we use the above diagram to feed back and make changes to the terms. We keep making shifts until there are n_2 or less terms, $c_0, c_2, \ldots, c_{n_2-1}$, left.

What is left is the result of the *Euclidean algorithm*. As an exercise, the reader is asked to set up an LFSR over F_3 using $f_2(x) = 1 - x + x^2 - x^3$.

2.5.2. *Ideals*

In the ring theoretical coding theory, the word space will be $\mathbf{R}/\mathbf{I}$, where $\mathbf{R} = \mathbf{F}[x]$, and $\mathbf{I}$ an ideal of $\mathbf{R}$. The code space will be $\mathbf{J}/\mathbf{I}$, where $\mathbf{J} \supset \mathbf{I}$ an ideal. We have the following.

Proposition 2.20. *Every ideal* $\mathbf{I}$ *of* $\mathbf{R} = \mathbf{F}[x]$ *is of the form* $(g(x))$.

Proof. We may use the preceding proposition to construct a generator $g(x)$. For an existence proof, note that the zero ideal is generated by 0, otherwise, suppose $\mathbf{I}$ is non-zero. Let $g(x)$ be a non-zero polynomial in $\mathbf{I}$ with the smallest degree. Then $g(x)$ is a generator by the Euclidean algorithm.
□

Corollary 2.21. *The ring* $\mathbf{F}[x]/\mathbf{I}$ *is a principal ideal ring.*

Proof. Let $\bar{\mathbf{J}}$ be an ideal in $\mathbf{R}/\mathbf{I}$. Let $\mathbf{J}'$ be the pre-image of $\bar{\mathbf{J}}$ in $\mathbf{R}$. Then, $\mathbf{J}'$ is clearly an ideal in $\mathbf{R}$. Let $g(x)$ be its generator. Then clearly, $\overline{g(x)}$ generates $\bar{\mathbf{J}}$.
□

The following proposition is the Lagrange interpolation theorem which can be considered as a special case of the *Chinese remainder theorem*.

Proposition 2.22. *Let* $\{\beta_i\}_1^n$, $\{\alpha_i\}_1^n$ *be elements in* $\mathbf{F}$ *such that all* β_i's *are distinct. Then, there is an unique polynomial* $f(x)$ *of degree at most* $n - 1$ *such that* $f(\beta_i) = \alpha_i \ \forall \ i$. *Moreover,* $f(x)$ *is of the following form:*

$$f(x) = \sum_i \alpha_i \frac{\prod_{j \neq i}(x - \beta_j)}{\prod_{j \neq i}(\beta_i - \beta_j)}.$$

Proof. It is easy to see that the $f(x)$ defined above satisfies the requirements of the proposition. Suppose $g(x)$ is another. Then, we have $f(x) - g(x)$ with n roots β_i and of degree at most $n - 1$. Therefore, $f(x) - g(x) = 0$ or $f(x) = g(x)$.
□

Corollary 2.23. *Let* $\{\beta_i\}$ *be a set of n distinct numbers. Let* $\mathbf{P}_n$ *be the set of all polynomials of degrees* $\leq n-1$. *The* $\mathbf{P}_n$ *is generated by* $\{\prod_{j \neq i}(x - \beta_j)\}$ *as a vector space over* $\mathbf{F}$. ∎

We make the following simple generalization of the above Lagrange interpolation theorem.

Proposition 2.24. *Let* $\{\beta_i\}_1^m$, $\{\alpha_i\}_1^n$ *be elements in* **F** *where* $m \leq n$ *and all* β_i's *are distinct and* α_i's *may not be all distinct. Furthermore, we have positive integers* m_i *(i.e., the multiplicities) such that* $\sum_{i=1}^m m_i = n$. *Then, there is a unique polynomial* $f(x)$ *of degree at most* $n - 1$ *such that* $f(x) - \alpha_i = (x - \beta_i)^{m_i} f_i(x) \forall\, i$ *for some suitable polynomials* $f_i(x)$, *with* $f_i(\beta_i) \neq 0$.

Moreover, there are polynomials $g_i(x)$ *with* $\deg(g_i(x)) \leq (m_i - 1)$ *that satisfy*

$$\sum_i g_i(x) \prod_{j \neq i} (x - \beta_j)^{m_j} = 1$$

such that $f(x)$ *is of the following form:*

$$f(x) = \sum_i \alpha_i g_i(x) \prod_{j \neq i} (x - \beta_j)^{m_j}.$$

Proof. First, we prove the equation

$$\sum_i g_i(x) \prod_{j \neq i} (x - \beta_j)^{m_j} = 1.$$

We shall use the proof of the Chinese remainder theorem. Since the polynomials $\prod_{j \neq i} (x - \beta_j)^{m_j}$'s are co-prime, they generate the unit ideal, i.e., we have

$$\sum_i h_i(x) \prod_{j \neq i} (x - \beta_j)^{m_j} = 1$$

for some suitable $h_i(x)$. We may not take h_i to be g_i because the degree restriction on g_i may not be satisfied. Let $r = \max\{\deg(h_i(x)) - m_i\}$. If $r < 0$, then we just let $g_i(x) = h_i(x)$. Note that the degree restriction is satisfied. If $r \geq 0$, let $s =$ the number of $h_i(x)$ such that $\deg(h_i(x)) - m_i = r$. We make a double induction on r, s, i.e., we reduce the number s to 0, then r will automatically drop. When r drops to negative, then we find $g_i(x)$'s.

Note that we always have $s > 1$. Otherwise, $s = 1$, and there is a unique term of the highest degree, which cannot be canceled by any other term, and the above equation cannot be satisfied. Therefore, $s \geq 2$. Let us pick up any two terms of the highest degree, say corresponding to $h_i(x)$ and $h_j(x)$. We have $r = \deg\, h_i(x) - m_i = \deg\, h_j(x) - m_j$. For any c, $h_i(x)$ and $h_j(x)$ can be replaced by $h_i(x) + cx^r(x - \beta_i)^{m_i}$ and $h_j(x) - cx^r(x - \beta_j)^{m_j}$, respectively, and the above equation is still satisfied. We may select c such that $\deg(h_i(x) + cx^r(x - \beta_i)^{m_i})$ is smaller. Thus, at least one term drops out from the collection of the highest terms. We reduce the number s at least by 1. When the number s drops to zero, then r must drop. So, we find $g_i(x)$'s by double induction.

Now, we have

$$\sum_i g_i(x) \prod_{j \neq i} (x - \beta_j)^{m_j} = 1 \quad \text{and}$$

$$\alpha_k \left(\sum_i g_i(x) \prod_{j \neq i} (x - \beta_j)^{m_j} \right) = \alpha_k.$$

Therefore,

$$\left(\sum_i \alpha_i g_i(x) \prod_{j \neq i} (x - \beta_j)^{m_j} \right) - \alpha_k$$

$$= \left(\sum_i \alpha_i g_i(x) \prod_{j \neq i} (x - \beta_j)^{m_j} \right) - a_k \left(\left(\sum_i g_i(x) \prod_{j \neq i} (x - \beta_j)^{m_j} \right) \right)$$

$$= \left(\sum_{i \neq k} \alpha_i g_i(x) \prod_{j \neq i} (x - \beta_j)^{m_j} \right) - a_k \left(\left(\sum_{i \neq k} g_i(x) \prod_{j \neq i} (x - \beta_j)^{m_j} \right) \right)$$

$$= (x - \beta_k)^{m_k} f_k(x).$$

It means that we may let $f(x) = \sum_i \alpha_i g_i(x) \prod_{j \neq i} (x - \beta_j)^{m_j}$, then we have $f(x) - \alpha_k = (x - \beta_k)^{m_k} \forall k$. Furthermore, if there is another polynomial $f^*(x)$ with the same properties of $f(x)$, namely

$$f(x) - \alpha_k = (x - \beta_k)^{m_k} f_k(x) \quad \text{and} \quad f^*(x) - \alpha_k = (x - \beta_k)^{m_k} f_k^*(x),$$

then $f(x) - f^*(x)$ will be divisible by $(x - \beta_i)^{m_i}$ for all i, i.e., $f(x) - f^*(x)$ will be divisible by $\prod_i (x - \beta)^{m_i}$ which has a degree n higher than $n - 1$. We conclude that $f(x) - f^*(x) = 0$. □

2.5.3. *A Ring Theoretical Presentation of a Hamming Code*

Before we study the abstract theory of rings further, let us study an example to illustrate the usage of ring theory to express the *Hamming code* and introduce the readers to the next level of coding theory. Recall the $[7, 4, 3]$

Hamming code **C** having check matrix H_1 as follows

$$H_1 = \begin{pmatrix} 1 & 1 & 0 \\ 0 & 1 & 1 \\ 1 & 1 & 1 \\ 1 & 0 & 1 \\ 1 & 0 & 0 \\ 0 & 1 & 0 \\ 0 & 0 & 1 \end{pmatrix}.$$

Let us permute the rows of H_1 to redefine H as follows:

$$H = \begin{pmatrix} 1 & 0 & 0 \\ 0 & 1 & 0 \\ 0 & 0 & 1 \\ 1 & 1 & 0 \\ 0 & 1 & 1 \\ 1 & 1 & 1 \\ 1 & 0 & 1 \end{pmatrix}.$$

Now, consider the ring $\mathbf{F}_2[x]/(1+x+x^3) = \mathbf{K}(= \mathbf{F}_{2^3})$. It is easy to see that the polynomial $g(x) = 1 + x + x^3$ is irreducible over $\mathbf{F}_2$ (for instance, if $g(x)$ can be factored, then one of the factor must be linear, so it has a root in $\mathbf{F}_2$. However, we have $g(0) \neq 0$ and $g(1) \neq 0$, then $g(x)$ cannot be factored). So, $\mathbf{K}$ is a field and isomorphic to $\mathbf{F}_2^3$ as vector spaces. If β is the image of x under the quotient mapping, then as a vector space we have

$$\mathbf{K} = \{ a_1\beta^0 + a_2\beta^1 + a_3\beta^2 : a_i \in \mathbf{F}_2 \}.$$

We can drop the powers of β and write rows of the coefficients $[a_1 a_2 a_3]$ as an element in the vector space $\mathbf{K}$ $(= \mathbf{F}_{2^3})$ as

$$\mathbf{K} = \{ [000], [100], [010], [001], [110], [011], [111], [101] \}.$$

On the other hand, $K \backslash \{0\}$ is a cyclic group multiplicatively, while we may write them as element in a field $\mathbf{K} = \{ 0, 1, \beta, \beta^2, \beta^3, \beta^4, \beta^5, \beta^6 \}$, where $\beta^7 = 1$. The interesting thing about these two representations is that the ordering of elements are identical (this is accidental). For instance, the sixth element in the field representation is β^4, while

$$x^4 \equiv x + x^2 \quad \mod \ (1 + x + x^3).$$

Therefore,

$$\beta^4 = 0 + 1\beta + 1\beta^2,$$

and β^4 corresponds to $[011]$, which is the sixth element in the vector space representation of $\mathbf{K}(= \mathbf{F}_{2^3})$.

We may write $H = (1\beta\beta^2\beta^3\beta^4\beta^5\beta^6)^T$ explicitly as

$$
H = \begin{bmatrix} 1 \\ \beta \\ \beta^2 \\ \beta^3 \\ \beta^4 \\ \beta^5 \\ \beta^6 \end{bmatrix} = \begin{bmatrix} 1 & 0 & 0 \\ 0 & 1 & 0 \\ 0 & 0 & 1 \\ 1 & 1 & 0 \\ 0 & 1 & 1 \\ 1 & 1 & 1 \\ 1 & 0 & 1 \end{bmatrix}.
$$

We define Hamming code space as all words $a = [a_1a_2a_3a_4a_5a_6a_7]$ such that $a \times H = 0$, i.e., $\sum_1^7 a_i\beta^{i-1} = 0$, where $a_i \in \mathbf{F_2}$. Alternatively, we may write each word as a polynomial $a(x) = \sum a_i x^{i-1}$ such that deg $a(x) \leq 6$, and $a(x)$ is a code word iff $a(\beta) = 0 \in \mathbf{F}_{2^3}$. At this point, we may further involve some known mathematical techniques to handle the coding problem.

The condition $a(\beta) = 0 \in \mathbf{F}_{2^3}$ implies $a(x) = (1 + x + x^3) \times m(x)$. Therefore, deg $m(x) \leq 3$ and $m(x) = m_1 + m_2x + m_3x^2 + m_4x^3$, and all $m(x)$ form the message space. The relation of $a(x)$ and $m(x)$ is as follows:

$$
\begin{aligned}
a(x) &= a_1 + a_2x + a_3x^2 + a_4x^3 + a_5x^4 + a_6x^5 + a_7x^6 \\
&= (1 + x + x^3)m(x) \\
&= (1 + x + x^3)(m_1 + m_2x + m_3x^2 + m_4x^3) \\
&= m_1 + (m_1 + m_2)x + (m_2 + m_3)x^2 + (m_1 + m_3 + m_4)x^3 \\
&\quad + (m_2 + m_4)x^4 + m_3x^5 + m_4x^6.
\end{aligned}
$$

We have the following *parametric form* of the code space with m_i's as the parameters:

$$
\begin{aligned}
& a_1 = m_1, \ a_2 = m_1 + m_2, \\
& a_3 = m_2 + m_3, \ a_4 = m_1 + m_3 + m_4, \qquad (3) \\
& a_5 = m_2 + m_4, \ a_6 = m_3, \\
& a_7 = m_4.
\end{aligned}
$$

After we eliminate the parameters m_i's, we have the following system of defining equations:

$$
\begin{aligned}
& a_1 = a_4 + a_6 + a_7, \\
& a_2 = a_4 + a_5 + a_6, \qquad (4) \\
& a_3 = a_5 + a_6 + a_7.
\end{aligned}
$$

Therefore, we have $[a_4a_5a_6a_7] \times G = [a_4a_5a_6a_7a_1a_2a_3]$, where G is the following matrix:

$$G = \begin{pmatrix} 1 & 0 & 0 & 0 & 1 & 1 & 0 \\ 0 & 1 & 0 & 0 & 0 & 1 & 1 \\ 0 & 0 & 1 & 0 & 1 & 1 & 1 \\ 0 & 0 & 0 & 1 & 1 & 0 & 1 \end{pmatrix}.$$

Note that G is the same generator matrix used in the introduction.

We can stay within the ring of polynomial $\mathbf{F}_2[x]$ by starting with message $[m_1 \ldots m_4]$, forming the polynomial $m(x) = \sum_1^4 m_i x^{i-1}$ with the code word $a(x)$ as

$$a(x) = (1 + x + x^3)m(x) = \sum_1^7 a_i x^{i-1}$$

and to send the encoded message as the coefficients $[a_1 \ldots a_7]$. The receiver will check the received $[a'_1 \ldots a'_7]$ and compute $a'(\beta) = \sum_1^7 a'_i \beta^{i-1}$.

Assume that there is at most one error in the received word. If the result of the preceding calculation is $a'(\beta) = 0$, then there is no error. Otherwise, $a'(\beta) = \beta^j \neq 0$, and there is an error at the $(j+1)$th spot of $[a'_1 \cdots a'_7]$. After correcting the error and recovering the original polynomial $a(x)$, one has the polynomial $m(x) = a(x)/(1 + x + x^3)$ and the original message $[m_1 \cdots m_4]$. We call the polynomial $g(x) = 1 + x + x^3$ the *generator polynomial* and the polynomial $h(x) = (1 + x^7)/g(x) = 1 + x + x^2 + x^4$ the *check polynomial*. Note that $a(x)$ is a code word iff $a(\beta) = 0$ iff $a(x) = c(x)g(x)$ iff $(1 + x^7)|a(x)h(x)$.

The decoder works perfectly if there is no error or if there is one error. However, if there more than one error, and (1) if r happens to be a code word, then the decoder will treat it as the original code word, or (2) if r is not a code word, then the decoder will replace it by a wrong code word.

The above example illustrates that Hamming codes can be discussed purely in terms of polynomial rings. It broadens our horizons. This is the next level of development in coding theory. Before we continue our discussion of $\mathbf{F}_q[x]$, we must have some understanding of the finite field $\mathbf{F}_q$.

Exercises

(1) Prove the Chinese remainder theorem for $\mathbf{K}[x]$, and use it to prove the Lagrange interpolation theorem.

(2) Find the total quotient ring of $\mathbb{Z}/4\mathbb{Z}$.

(3) Let $\chi_\gamma(x)$ be the characteristic function of $\gamma \in \mathbf{K}^n$, where $\mathbf{K}$ is a finite field, i.e., $\chi_\gamma(\gamma) = 1$ and $\chi_\gamma(a) = 0$ for all $a \neq \gamma$. Show that $\chi_\gamma(x)$ can be written as a polynomial in n variables.

(4) Let π be any $\mathbf{K}$-valued function on $\mathbf{K}^n$, where $\mathbf{K}$ is a finite field. Show that π can be written as a polynomial in n variables.

(5) Find two distinct polynomials $f(x), g(x) \in \mathbf{F}_p[x]$ such that $f(\gamma) = g(\gamma)$ for all $\gamma \in \mathbf{F}_p$.

2.6. Separability

We have the following useful definition.

Definition 2.25. A polynomial $f(x) \in \mathbf{K}[x]$ is said to be separable if it has no multiple root in an algebraic closure Ω of $\mathbf{K}$.   ∎

Proposition 2.26. *Over a finite field $\mathbf{F}_q$, every irreducible polynomial is separable.*

Proof. Let $f(x)$ be an irreducible polynomial. If it has a multiple root, then the root must be a root of $f'(x)$. It means that $f(x)$ and $f'(x)$ must have a common factor. With the assumption of irreducibility of $f(x)$, this can happen only if $f'(x)$ is the zero polynomial, i.e., $f(x) = g(x^p)$ for some suitable $g(x)$. Since a finite field is perfect, every coefficient of $g(x)$ is a pth power of some other element. It is easy to see that $f(x) = g(x^p) = h(x)^p$ for some suitable $h(x)$. This contradicts the irreducibility of $f(x)$.   □

Let $q = p^\ell$. We wish to count the number of monic irreducible polynomials of degree r in $\mathbf{F}_{q^m}[x]$. Let the number be $\mathbf{I}_r$.

Proposition 2.27. *Let s be any positive integer. We have*

$$q^{ms} = \sum_{r \mid s} r \mathbf{I}_r.$$

Proof. Let us fix an algebraic closure Ω of $\mathbf{F}_{q^m}$. Let $\alpha \in \mathbf{F}_{q^m}$ satisfying a monic irreducible polynomial $f_\alpha(x) \in \mathbf{F}_{q^m}[x]$ of degree, say r. From finite-field theory, we have $r \mid s$. From Proposition 2.3, we know that $f_\alpha(x)$ is separable. Any root of $f_\alpha(x)$ will generate a field extension over $\mathbf{F}_{q^m}$ of degree r. It follows from **Proposition 1.34** that this field extension is contained in $\mathbf{F}_{q^{ms}}$. We may group all roots (there are r of them) of $f_\alpha(x)$ together. Our proposition follows easily.   □

Now we use the well-known Möbius inversion formula in number theory to find $\mathbf{I}_r$. For this purpose, we have the following definition.

Definition 2.28. The Möbius function μ is defined as

$$\mu(n) = \begin{cases} 1, & \text{if } n=1, \\ (-1)^k, & \text{if } n \text{ is a product of } k \text{ distinct primes,} \\ 0, & \text{otherwise.} \end{cases}$$ ∎

The following theorem is well known.

Theorem 2.29 (Möbius inversion formula). *We have*

$$\text{If } g(n) = \sum_{d|n} f(d), \ \forall n$$

$$\text{then } f(n) = \sum_{d|n} \mu(d) g\left(\frac{n}{d}\right) \ \forall n.$$

Proof. (1) The Möbius function $\mu(n)$ satisfies

$$\sum_{d|n} \mu(d) = \begin{cases} 1, & \text{if } n=1, \\ 0, & \text{otherwise.} \end{cases}$$

(2) We have the following:

$$\sum_{d|n} \mu(d) g\left(\frac{n}{d}\right) = \sum_{d|n} \mu\left(\frac{n}{d}\right) g(d)$$

$$= \sum_{d|n} \mu\left(\frac{n}{d}\right) \sum_{k|d}(f(k)) = \sum_{k|n} \sum_{d|(n/k)} \mu\left(\frac{n}{kd}\right) f(k)$$

$$= \sum_{k|n} \sum_{d|(n/k)} \mu(d) f(k) = \sum_{k|n} \sum_{d|(n/k)} \mu(d) f(k)$$

$$= f(n).$$ □

Remark: The converse is also true. ∎

Recall that I_r is the number of monic irreducible polynomials of degree r in $F_{q^m}[x]$. It leads to the following proposition.

Proposition 2.30. *If $r \geq 1$, then $I_r > 0$. Let r be a prime. If r is 2, then* $r\mathbf{I}_r = (q^{mr} - q^m) > 0$.
Otherwise, $r\mathbf{I}_r = \sum_{d|r} \mu(d) q^{mr/d} > q^{mr}(1 - q^{-rm/2+1}) > 0$.

Proof. It is clear that every linear polynomial is irreducible. We have $r = 1$ which implies that $I_r > 0$. If r is 2, then $r\mathbf{I}_r = (q^{mr} - q^m) > 0$.

Let us assume that $r > 2$ and is a prime. It follows from Proposition 2.27 and Theorem 2.29 that

$$r\mathbf{I}_r = \sum_{d|r} \mu(d) q^{mr/d}.$$

Then, we have $I_r > 0$. In general, we have $r \geq 3$ and

$$\begin{aligned}
r\mathbf{I}_r &= \sum_{d|r} \mu(d) q^{mr/d} \\
&> \left(q^{mr} - \sum_{i \geq 2} q^{md/i} \right) \\
&> \left(q^{mr} - q^{mr/2+1} \right) \\
&= q^{rm} \left(1 - q^{-rm/2+1} \right) \\
&> 0
\end{aligned}$$

and our proposition is proved. $\qquad\qquad\qquad\qquad\qquad\qquad\qquad\square$

The above proposition shows the existence of finite field F_{p^m} for any m in a fixed algebraic closure Ω. Each finite field F_{p^m} exists uniquely. We have the following diagram, where lines indicate inclusive relation.

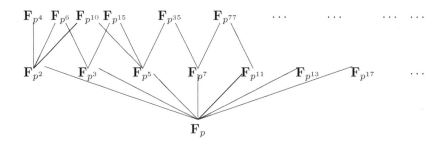

Figure 2.1.

2.7. Power Series Rings F[[x]] and Fields of Meromorphic Functions F((x))

We shall consider formal power series in this section. First, the decoding process of codes on $\mathbf{F}[x]$ depend on some properties in this section (see Proposition 2.36). Second, the important concept of *residue* in a Riemann surface (or a smooth algebraic curve), which is all important in the *Geometric Goppa Codes*, can be computed (see Proposition 4.42) with the help of the materials in this section.

Let us consider the expression $f(x) = \sum_{i=0}^{\infty} a_i x^i$, where a_i's are coefficients in the field $\mathbf{F}$. In the past, analysts considered the problem of evaluating the expression for $x = b$ and deducing many concepts of convergence, divergence, etc. If a power series is divergent, then we are likely to disregard it from the point of analysis. However, algebraically, if we do not evaluate them (other than the trivial evaluation at $x = 0$), then we may simply treat them as algebraic items, and they play their roles. Henceforth, expressions of the form

$$f(x) = \sum_{i=0}^{\infty} a_i x^i$$

will be called *formal power series* or *power series*, and expressions of the form

$$f(x) = \sum_{i=-m}^{\infty} a_i x^i$$

will be called *formal meromorphic functions* or *meromorphic functions*. All formal power series with coefficients in $\mathbf{F}$ will be denoted by $\mathbf{F}[[x]]$. All formal meromorphic functions with coefficients in $\mathbf{F}$ will be denoted by $\mathbf{F}((x))$.

Let $f(x) = \sum_j a_j x^j$, $g(x) = \sum_k b_k x^k \in \mathbf{F}((x))$. We define, as usual,

$$f(x) \cdot g(x) = \sum_i \left(\sum_{j+k=i} a_j b_k \right) x^i$$

$$f(x) + g(x) = \sum_i (a_i + b_i) x^i.$$

We have the following proposition.

Proposition 2.31. *The sets* $\mathbf{F}[[x]] \subset \mathbf{F}((x))$ *are closed under the usual addition and multiplication.*

Proof. The proposition follows by routine checking. □

Definition 2.32. Let $f(x) = \sum_i c_i x^i$ be a formal power series or a formal meromorphic function. We define the *order* of $f(x)$, in symbol $\mathrm{ord}(f(x))$, as

$$\mathrm{ord}(f(x)) = \begin{cases} \min\{i : c_i \neq 0\} & \text{if } f(x) \neq 0, \\ \infty & \text{if } f(x) = 0. \end{cases} \quad ■$$

We have the following proposition.

Proposition 2.33. *We have*

$$\mathrm{ord}(f(x)g(x)) = \mathrm{ord}(f(x)) + \mathrm{ord}(g(x)),$$

$$\mathrm{ord}(f(x) + g(x)) \geq \min(\mathrm{ord}(f(x)), \mathrm{ord}(g(x))).$$

Proof. The proof is a routine check. □

Proposition 2.34. *We have* $f(x)$ *which is a unit in* $\mathbf{F}[[x]]$ *(i.e.,* $f(x)^{-1}$ *exists)* $\Longleftrightarrow f(0) \neq 0$.

Proof. ($\Longrightarrow$) If $f(x)$ is a unit, then there exists $g(x)$ such that $f(x)g(x) = 1$. Then, we have

$$\mathrm{ord}(f(x)) + \mathrm{ord}(g(x)) = 0.$$

Therefore, $\mathrm{ord}(f(x)) = 0$ and $f(0) \neq 0$.

($\Longleftarrow$) If $f(0) = a \neq 0$, then $a^{-1}f(0) = 1$. We may write $a^{-1}f(x)$ as $1 - g(x)$ with $\mathrm{ord}(g(x)) \geq 1$. We have

$$(a^{-1}f(x))^{-1} = 1 + \sum_{i=1}^{\infty} g(x)^i.$$

It is easy to see that when we collect coefficients of like degrees in x on the right-hand side and add them up, we only have finite sums. Therefore, the right-hand side is a power series. It is then trivial to see that the inverse of $f(x)(=a^{-1}(a^{-1}f(x))^{-1})$ exists. □

Note that a ring $\mathbf{R}$ is said to be an **integral domain** if $f \cdot g = 0$, then either $f = 0$ or $g = 0$. We have the following.

Proposition 2.35. *Given any field* $\mathbf{F}$, *the set* $\mathbf{F}[[x]]$ *is an integral domain and* $\mathbf{F}((x))$ *is field.*

Proof. The proof is a routine check. □

Let us use the above computations for polynomials. Let $f(x) = \sum_{i=0}^{m} a_i x^i$, $g(x) = \sum_{i=0}^{n} b_i x^i$ be two polynomials of degrees m, n, respectively. For simplicity, let us assume that $b_0 = g(0) \neq 0$. Then, $\frac{f(x)}{g(x)} \in \mathbf{F}[[x]]$. We have

$$\frac{f(x)}{g(x)} = \sum_{i=0} c_i x^i.$$

The total number of coefficients of $\frac{f(x)}{g(x)}$ is $(m + 1 + n + 1)$. Let us compute the number of independent coefficients. Furthermore, we may change $f(x), g(x)$ by a common multiple of non-zero constant without changing $\frac{f(x)}{g(x)}$. Therefore, the number of independent coefficients should be $(m + n + 1)$. It is natural to ask if the first $(m + n + 1)$ coefficients $c_0, c_1, \ldots, c_{m+n}$ of the above power-series expression uniquely determine the rational function $\frac{f(x)}{g(x)}$. We have the following natural proposition which is significant for decoding purposes.

Proposition 2.36. *Given any polynomial* $h(x)$ *of degree* $m + n$, *there is at most one rational function* $\frac{f(x)}{g(x)}$, *where* $f(x), g(x)$ *are of degrees at most* m, n, *respectively, with* $g(0) \neq 0$ *such that in the following equation in* $\mathbf{F}[[x]]$:

$$\text{ord}\left(\frac{f(x)}{g(x)} - h(x)\right) \geq m + n + 1 \tag{1}$$

Furthermore, (1) if such a rational function $\frac{f(x)}{g(x)}$ *exists, and there exists another pair of polynomials* $(f_1(x), f_2(x))$ *with* $f_2(0) \neq 0$, *with* $f_1(x), g_1(x)$ *being of degrees at most* $m - s, n - s$, *respectively, for some non-negative integer* s, *then* $\frac{f_1(x)}{g_1(x)} = \frac{f(x)}{g(x)} \iff$

$$\text{ord}\left(\frac{f_1(x)}{g_1(x)} - h(x)\right) \geq m + n - s + 1. \tag{2}$$

(2) If a factor $a(x)$ *of degree* u *of* $g(x)$ *is known, then we have*

$$\text{ord}\left(\frac{f(x)}{g(x)} - h(x)\right) \geq m + n - u + 1. \tag{3}$$

Proof. Let us show the uniqueness. If $\frac{f^*(x)}{g^*(x)}$ is another pair of polynomials, where $f^*(x), g^*(x)$ are of degrees at most m, n, respectively, with $g^*(0) \neq 0$ and

$$\mathrm{ord}\left(\frac{f^*(x)}{g^*(x)} - h(x)\right) \geq m + n + 1.$$

Then, we have

$$\mathrm{ord}\left(\frac{f(x)}{g(x)} - \frac{f^*(x)}{g^*(x)}\right) \geq m + n + 1.$$

On the other hand, we have

$$\frac{f(x)}{g(x)} - \frac{f^*(x)}{g^*(x)} = \frac{f(x)g^*(x) - f^*(x)g(x)}{g(x)g^*(x)}.$$

Note that $g(0)g^*(0) \neq 0$. We conclude

$$\mathrm{ord}(f(x)g^*(x) - f^*(x)g(x)) \geq m + n + 1.$$

Note that $f(x)g^*(x) - f^*(x)g(x)$ is a polynomial of degree at most $m + n$, if it is not zero. Note that for any non-zero polynomial, its degree is bigger or equal to its order. Hence. it must be zero. We have

$$\frac{f(x)}{g(x)} = \frac{f^*(x)}{g^*(x)}.$$

Thus, we have the uniqueness. The proof of the part (1) of the proposition is as follows. For $\Longleftarrow$, we follow almost verbatim as above. For $\Longrightarrow$, we have

$$\mathrm{ord}\left(\frac{f_1(x)}{g_1(x)} - h(x)\right) \geq m + +n + 1 \geq m + n - s + 1.$$

Hence, the proof of (1) is complete. The proof of the part (2) of the proposition is similar to the above and left to the reader. $\square$

Example 4: The preceding proposition claims that there is **at most** one rational function $\frac{f(x)}{g(x)}$ that satisfies all conditions. We give the following example to show that there may not exist a rational function $\frac{f(x)}{g(x)}$ with the said property. For instance, let $h(x) = 1 + x^4$, $m = n = 2$, then there is no $\frac{f(x)}{g(x)}$ with $\deg(f(x)) \leq 2$ and $\deg(g(x)) \leq 2$ such that

$$\mathrm{ord}\left(\frac{f(x)}{g(x)} - h(x)\right) \geq m + n + 1 = 5.$$

Suppose the contrary, there are polynomials $f(x), g(x), h(x)$ that satisfied the above equation. Let us multiply by $g(x)$ the equation $\frac{f(x)}{g(x)} = h(x) mod(x^5)$; it will produce

$$f(x) = g(x) + g(0)x^4 \quad \mod \ (x^5).$$

Since we have the restrictions on the degrees of $f(x), g(x)$, we must have $f(x) = g(x)$, which do not satisfy all numerical conditions. ∎

Exercises

(1) Factor $x^{16} - x$ over $\mathbf{F}_2$ into a product of irreducible polynomials.
(2) Given a finite field $\mathbf{F}_p$, show that the following polynomial $f(x)$ is separable:

$$f(x) = 1 + \cdots + \frac{x^i}{i!} + \cdots + \frac{x^{p-1}}{(p-1)!}.$$

(3) Let $\mathbf{K}$ be a field with $\mathrm{ch}(\mathbf{K}) = p$ positive. Show that $x^p - x - a$ is either irreducible or splits completely into linear factors, where $a \in \mathbf{K}$.
(4) Let $z = x^2 + x^3$. Show that $[\mathbf{F}_2((x)) : \mathbf{F}_2((z))] = 2$.
(5) Let $\mathbf{K}$ be a field of characteristic p and x, y be symbols. Show that $\mathbf{K}(x, y)$ is inseparable over $\mathbf{K}(x^p, y^p)$.

Chapter 3

Ring Codes

From now on, we shall fix the ground field $\mathbf{F}_q$. A linear code is defined by a subspace $\mathbf{C}$ of an n-dimensional vector space $\mathbf{F}_q^n$. Note that for the purpose of decoding, we shall have more (algebraic) relations between the vectors in the *words space*. Naturally, there shall be multiplicative relations between the vectors in the *words space*. It means that we shall consider *ring* structures. A simple ring is $\mathbf{F}_q[x]/(f(x))$. It is easy to see that $\mathbf{F}_q^n$ can be represented by $\mathbf{F}_q[x]/(f(x))$ for any polynomial $f(x)$ of degree n. In this representation, we have the multiplicative structure of a ring other than the additive structure of a vector space. We shall study the concepts of coding theory in the context of the ring $\mathbf{F}_q[x]/(f(x))$. To make our work easier, we shall later select a *good* $f(x)$ for coding purposes. We have the following definition to begin with.

Definition 3.1. Given any polynomial $h(x)$, the Hamming weight of $h(x)$ mod $(f(x))$ is to consider $\bar{h}(x) \in \mathbf{F}_q[x]/(f(x))$ as $\bar{h}(x) = \sum_{i=0}^{n-1} c_i x^i$, where $n = deg(f(x))$, and the Hamming weight of $h(x)$ mod $(f(x))$ is the Hamming weight of $[c_0, c_1, \ldots, c_{n-1}]$. Note that the preceding expression $\bar{h}(x)$ is unique in any residue class with the degree less then n. ∎

There is a convenient way to consider only **code subspaces** defined by ideals $\bar{\mathbf{I}}$ in $\mathbf{F}_q[x]/(f(x))$, and in this case, the elements in $\bar{\mathbf{I}}$ are called code words. Note that $\mathbf{F}_q[\mathbf{x}]$ is a principal ideal ring, for instance let $\bar{\mathbf{I}}$ be any ideal in $\mathbf{F}_q[x]/(f(x))$ and $\mathbf{I}$ be its pull back in $\mathbf{F}_q[x]$, i.e., $\mathbf{I} = \{g(x) : \bar{g}(x) \in \bar{\mathbf{I}}$, then $\mathbf{I}$ is a principal ideal. We write $\mathbf{I} = (h(x))$. It is clear that $\bar{\mathbf{I}} = (\bar{h}(x))$. So we conclude that $\mathbf{F}_q[x]/(f(x))$ is a principal ideal ring.

Since an ideal $\mathbf{I} = (f(x), h(x))$ is generated by the greatest common divisor $h^*(x)$ of $f(x), h(x)$, we must have that $h^*(x)$ is a factor of $f(x)$. It is easy to see that any ideal $\bar{\mathbf{I}} \neq (0)$ of $\mathbf{F}_q[x]/(f(x))$ is of the form $(\bar{h}(x))$, where $h(x)$ is a factor of $f(x)$ in $\mathbf{F}_q[x]$. If $h(x) = 1$, then $\bar{\mathbf{I}} =$ the whole ring, or the code space = the word space; it is a non-interesting case. The interesting case is that $h(x)$ is a non-unite factor of $f(x)$ in $\mathbf{F}_q[x]$.

If we take $f(x)$ to be $x^n - 1$, then the coding theory will be very *interesting*. In general we have a *cyclic* code in the following sense.

Definition 3.2. A code space $\mathbf{C}$ of length n is said to be *cyclic* iff $[a_0 a_1 \ldots a_{n-1}] \in \mathbf{C} \Leftrightarrow [a_{n-1} a_0 \ldots a_{n-2}] \in \mathbf{C}$.   ∎

A cyclic code is rich in algebraic structures. We have the following equivalent characterizations.

Proposition 3.3. *Any ideal $\bar{\mathbf{I}} \neq (0)$ in $\mathbf{F}_q[x]/(x^n - 1)$ defines a cyclic code. Conversely, any cyclic code of length n can be represented by an ideal $\bar{\mathbf{I}}$ in $\mathbf{F}_q[x]/(x^n - 1)$.*

Proof. If $\bar{\mathbf{I}} = (1)$, then the code space is the whole word space, and the code space is cyclic. Let an ideal $\bar{\mathbf{I}} \neq (1)$ in $\mathbf{F}_q[x]/(x^n - 1)$ and a code word $c(x) = c_0 + c_1 x + \cdots + c_{n-1} x^{n-1} \in \bar{\mathbf{I}}$. Then $xc(x) \in \bar{\mathbf{I}}$ and $xc(x) = c_{n-1} + c_0 x + \cdots + c_{n-2} x^{n-1}$ since $x^n = 1$, and the code space is cyclic.

Conversely, if $\mathbf{C}$ is cyclic of length n, then $c(x) = c_0 + c_1 x + \cdots + c_{n-1} x^{n-1}$ forms a set $\bar{\mathbf{I}}$ for all $[c_0 c_1 \ldots c_{n-1}] \in \mathbf{C}$. The cyclic property of $\mathbf{C}$ implies $xc(x) \in \bar{\mathbf{I}}$; hence, $x^\ell c(x) \in \bar{\mathbf{I}}$. It is easy to see that $\bar{\mathbf{I}}$ is an ideal in $\mathbf{F}_q[x]/(x^n - 1)$, and $\mathbf{C}$ can be represented by $\bar{\mathbf{I}}$.   □

The case $p \mid n$ in the above is not interesting because then we have $x^n - 1 = (x^{n/p} - 1)^p$; therefore, $(x^{n/p} - 1)$ is a non-zero word, while its pth power is zero. This is against intuition and is a *degenerate case*. From now on, we only discuss the *non-degenerate cases* and **assume** that n, p are co-prime. Note that every non-zero ideal $\bar{\mathbf{I}}$ can be generated by a factor of $x^n - 1$, and they are all precisely meaningful cyclic codes.

Example 1: Let us consider *Hamming code* $[7, 4, 3]$ as in Section 2.5. It is a cyclic code and can be represented by the ideal $(x^3 + x + 1)$ in $\mathbf{F}_2[x]/(x^7 + 1) = \mathbf{F}_2[x]/(x^7 - 1)$.   ∎

Definition 3.4. The ideal **I** (or a cyclic code **C**) is generated by a monic polynomial $g(x)$, which is a factor of $x^n - 1$. It is called the **generator polynomial** of the cyclic code $\mathbf{C} = (\bar{g}(x))$, and the polynomial $h(x) = (x^n - 1)/g(x)$ is called the **check polynomial** of **C**. ∎

Let Ω be an algebraic closure of $\mathbf{F}_q$. We denote $f(x) = x^n - 1$. Under our assumption n, p are co-prime, we have

$$f'(x) = nx^{n-1},$$

which has only 0 as a root; furthermore, 0 is not a root of $f(x)$. Therefore, the polynomial $x^n - 1$ will have no multiple roots in Ω by the derivative test (cf. Proposition 2.7). Therefore, its divisor $g(x)$, the *generator polynomial*, will have no multiple roots. Let $\{\gamma_j\}$ be the set of all roots of $g(x)$ in Ω. Then, it is clear that $r(x) \in$ the ideal $(g(x))$ iff $r(\gamma_j) = 0$ for all j. Thus, the ideal (or subspace **C**) can be defined by $\{r(x) : r(\gamma_j) = 0 \mod (x^n - 1), \forall j\}$. In fact, we may not have to take all roots of $g(x)$ in Ω; a partial set of roots may be enough.

Proposition 3.5. *Given any set $\{\gamma_j\} \subset \Omega$ such that all γ_j are roots of $x^n - 1$ for a fixed n, let $g(x)$ be the least-common multiple of all monic irreducible polynomials satisfied by $\{\gamma_j\}$ for all j. Then, $g(x)$ is the generator polynomial of a cyclic code **C** in $\mathbf{F}_q[x]/(x^n - 1)$.*

Proof. Let $g_i(x)$ be the monic irreducible polynomial satisfied by γ_i. Then, it is clear that $g_i(x) \mid x^n - 1$ for all i. Therefore, $x^n - 1$ is a common multiple of all $g_i(x)$. By definition, $g(x)$ is the least-common multiple and $g(x) \mid x^n - 1$. □

3.1. BCH Codes

Let us recall a significant property of the Hamming code as described by Proposition 1.21, which concludes that the Hamming distance of such a code is 3 by showing that in the matrix of all Hamming code words as column vectors, any two column vectors are distinct, while any three column vectors are linearly dependent. To study the Hamming distances of other code, we shall consider the well-known Verdermonde matrices whose column vectors are linearly independent. We have the following proposition.

Proposition 3.6. *Let $x_i \in \mathbf{F}$ and M be the following $n \times n$ matrix:*

$$M = \begin{pmatrix} 1 & 1 & 1 & 1 & 1 & \dots & 1 \\ x_1 & x_2 & x_3 & x_4 & x_5 & \dots & x_n \\ x_1^2 & x_2^2 & x_3^2 & x_4^2 & x_5^2 & \dots & x_n^2 \\ x_1^3 & x_2^3 & x_3^3 & x_4^3 & x_5^3 & \dots & x_n^3 \\ \dots & \dots & \dots & \dots & \dots & \dots & \dots \\ x_1^{n-1} & x_2^{n-1} & x_3^{n-1} & x_4^{n-1} & x_5^{n-1} & \dots & x_n^{n-1} \end{pmatrix}.$$

Then, we have

$$\det M = \prod_{i>j}(x_i - x_j).$$

Proof. First, we treat all x_i as symbols. Subtracting the first column from the second column, ..., the first column from the nth column , we may extract $x_i - x_1$ from the ith column, we conclude that $\prod_{i>1}(x_i - x_1) \mid \det(M)$. By symmetry, we conclude

$$\prod_{i>j}(x_i - x_j) \mid \det(M).$$

Since both sides are polynomials of degrees $n(n-1)/2$, they can only differ by a constant. Comparing the coefficients of $\prod_{i\geq 2} x_i^{i-1}$ on both sides, we conclude that they must be equal. Since the proposition is true for x_i as symbols, it must be true for x_i as elements of the field $\mathbf{F}$. $\quad\square$

Let us further consider the ring $\mathbf{F}_q[x]/(x^n - 1)$ with the usual assumption that n, p are co-prime. For the purpose of coding theory, we have the following proposition.

Proposition 3.7. *Let γ be a primitive nth root of unity (i.e., γ is a root of $x^n - 1$ and not a root of $x^s - 1$ for any $0 < s < n$) in an algebraic closure Ω of the finite field $\mathbf{F}$. Let $g(x)$ be the least-common multiple of all monic irreducible polynomials (i.e., satisfied by everyone) of $\gamma^\ell, \gamma^{\ell+1}, \dots, \gamma^{\ell+\delta-1}$, where ℓ, δ are some non-negative integers less than $n-1$. Let $\mathbf{C}$ be the cyclic code with $g(x)$ as the generator polynomial. Then, we have*

$$\min\{Hamming\ wt(a) : a \in \mathbf{C}, a \neq 0\} \geq \delta.$$

Proof. Recall that the code space is the ideal generated by $g(x)$. Hence, all code polynomial $c(x)$ must be satisfied by $\gamma^\ell, \gamma^{\ell+1}, \dots, \gamma^{\ell+\delta-1}$. Suppose that the proposition is false. Then, there is a non-zero code polynomial

$$c(x) = \sum_{i_j} c_{i_j} x^{i_j}$$

with s terms where $s < \delta$ such that $c(\gamma^\ell) = \cdots = c(\gamma^{\ell+\delta-1}) = 0$. Consider the following system of equations

$$\sum_{i_j} c_{i_j} \gamma^{\ell i_j} = 0,$$

$$\cdots \cdots$$

$$\sum_{i_j} c_{i_j} \gamma^{(\ell+\delta-1)i_j} = 0.$$

Among the above δ linear equations in fewer than δ variables c_{i_j}, we pick the first s so that the number of equations matches the number of variables. The coefficient matrix is the following:

$$N = \begin{pmatrix} \gamma^{\ell i_1} & \gamma^{(\ell)i_2} & \cdots & \gamma^{\ell i_s} \\ \gamma^{(\ell+1)i_1} & \gamma^{(\ell+1)i_2} & \cdots & \gamma^{(\ell+1)i_s} \\ \cdots & \cdots & \cdots \cdots \\ \gamma^{(\ell+s-1)i_1} & \gamma^{(\ell+s-1)i_2} & \cdots & \gamma^{(\ell+s-1)i_s} \end{pmatrix}.$$

It suffices to show that the matrix N is non-singular, then all c_{i_j} must be zero; this implies $c(x)$ is the zero polynomial. Contradiction!

Let us show that the matrix N is non-singular. Let us pull $\gamma^{\ell i_j}$ from the jth column. Then, we have the following matrix L:

$$L = \begin{pmatrix} 1 & 1 & \cdots & 1 \\ \gamma^{i_1} & \gamma^{i_2} & \cdots & \gamma^{i_s} \\ \cdots & \cdots & \cdots \cdots \\ \gamma^{i_1(s-1)} & \gamma^{i_2(s-1)} & \cdots & \gamma^{i_s(s-1)} \end{pmatrix},$$

which is a Verdermonde matrix of rank s with x_j replaced by γ^{i_j}. Since we have $\gamma^{i_j} - \gamma^{i_k} \neq 0$ with $i_j \neq i_k \leq n-1$, the matrices L and N are non-singular. Therefore, all $c_{i_j} = 0$, and $c(x)$ is a zero polynomial, contrary to our assumption that $c(x)$ is a non-zero polynomial. $\square$

The following cyclic codes were discovered by Bose and Ray-Chaudhuri (1960) and Hocquenghem (1959) and are known as *BCH codes*.

Definition 3.8. Let γ be a *primitive* nth root of unity in Ω. Let $g(x)$ be the least-common multiple of all monic irreducible polynomials of

$\gamma^{\ell}, \gamma^{\ell+1}, \ldots, \gamma^{\ell+\delta-1}$, where ℓ, δ are some non-negative integers less than $n - 1$. Then, the cyclic code $\mathbf{C}$ with $g(x)$ as the generator polynomial in $\mathbf{F}_q[x]/(x^n - 1)$ is called the *BCH code* of *designed distance* δ and *length n*. Usually, we take $\ell = 1$. If $n = q^s - 1$ (i.e., γ is a primitive element in $\mathbf{F}_{q^s}$), then it is called a *primitive BCH code*. ∎

We have the following corollary of the preceding proposition.

Corollary 3.9. *The Hamming distance of a BCH code is at least δ.*

Proof. It follows from the preceding proposition. □

The Hamming codes use the vector space structure of $\mathbf{F}_2^n$. The BCH codes use the *ring* structure of $\mathbf{F}_2[x]/(x^n - 1)$. Note that $\mathbf{F}_2[x]/(x^n - 1)$ is isomorphic to $\mathbf{F}_2^n$ as a vector space, and furthermore, it has a rich ring structure. The ring structure makes them better.

Example 2: Let us consider a BCH code $\mathbf{C}$ over $\mathbf{F}_{2^4}$. Let us consider α, β, where α satisfies $x^2 + x + 1 = 0$ over $\mathbf{F}_2$ and β satisfies $x^2 + x + \alpha = 0$ over $\mathbf{F}_{2^2}$. It is not hard to see that $\mathbf{F}_{2^2} = \mathbf{F}_2[\alpha]$ and $\mathbf{F}_{2^4} = \mathbf{F}_{2^2}[\beta] = \mathbf{F}_2[\beta]$. Let $\gamma = \beta$ satisfying the minimal equation $x^4 + x + 1 = 0$ over $\mathbf{F}_2$. Then, certainly $\gamma^{15} - 1 = 0$ and γ is a primitive 15th root of unity. Let us consider the BCH code determined by $\gamma, \gamma^2, \gamma^3, \gamma^4, \gamma^5$. Note that $\delta = 5$ in this case. By checking, one can conclude that $\gamma, \gamma^2, \gamma^4, \gamma^5$ satisfy the following equation:

$$1 + x + x^3 + x^4 + x^5 + x^7 + x^8 = 0,$$

and γ^3 satisfies the following equation:

$$1 + x + x^2 + x^3 + x^4 = 0.$$

So, the generator polynomial $g(x)$ is the product of the above two polynomials, which is

$$(1 + x + x^3 + x^4 + x^5 + x^7 + x^8)(1 + x + x^2 + x^3 + x^4)$$

$$= 1 + x^3 + x^6 + x^9 + x^{12}$$

and the check polynomial $h(x)$ is $(x^{15} - 1)/g(x) = 1 - x^3$. It is easy to see that the generator polynomial and the check polynomial multiply to $x^{15} - 1 = x^{15} + 1$. For this code, the later Example 5 shows that it can correct up to two errors and is not a Hamming code. ∎

3.2. Decoding a Primitive BCH Code

For a useful code, the decoding process is significant. **What do we mean by** *decoding* **is that there is an integer** t **such that given a received word** r, **the following are possible: (1) If there are less than or equal** t **errors, then the decoder will find the original code word. (2) If there more than** t **errors, then either we find a code word** c **within a distance** $\leq t$ **from the received word (in general, if we find a code word, then with a small probability,** c **may not be the originally sent code word)** or we return an *error* message to indicate that what was found is not even a code word, and hence there are more than t errors. There are several fast ways of decoding a primitive BCH code. Other than one presented in this section, there are other ways; for instance, Berlekamp's algorithm, which is at least equally fast (see Appendix D). The advantage of the method presented here is its mathematical simplicity.

Let us consider a primitive BCH code of length n over $\mathbf{F}_q$ with designed distance δ. Let $g(x)$ be the generating polynomial and $c(x) = c_0 + c_1 x + \cdots + c_{n-1} x^{n-1} = s(x) g(x)$ be a code word.

During the transmission of the code word, there might be errors and *erasures*, which means the positions of the erased data are known, while the precise data at those positions are unavailable. Let the received word be $\bar{r}(x) = \cdots + E_i x^i + \cdots$, where i's indicate the erasure positions and E_i indicate the apparent values. We shall treat the letter E_i as 0 and define $r(x) = \cdots + 0 x^i + \cdots$. Let the positions of erasures be a set N which is known. Let the number of erasures be u. Let the hypothetical error word $e(x)$ be defined by $c(x) = r(x) - e(x)$. Clearly, given $r(x)$, then $c(x)$ and $e(x)$ determine each other. In the following *flowchart*, given $r(x)$, the decoder will produce $e(x)$ if the number of errors and erasures are within a limit. Otherwise, it will return an *error* message.

We have the following Figure 3.1.

In the below Figure 3.1, upon receiving the received word, we let it pass the *check polyn*, which will use the agreed check polynomial $h(x)$ to check if it is a code word. If it passes, then it goes through the *pass* box, and it will be declared as a *code word*; in other words, it flows through the left half of *Flowchart 1*. If it fails the *check polyn*, then it goes through the right half of *Flowchart 1* and start with the *decoder*. The decoder[1] is based on the

---

[1] We follow the decoding process of SKHN [35]. Another slightly faster decoding process is the Berlekamp's algorithm [4] decoder. Please see Appendix D.

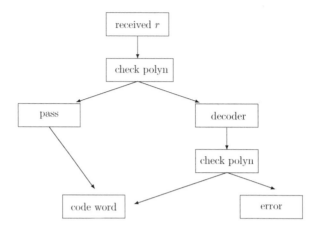

Figure 3.1.

assumption that the number of errors v and erasures u are limited by *the numerical condition $2v + u < \delta$*, and we shall use the theory of power series and the Euclidean algorithm. The *decoder* will produce an error word e (see Section 3.2) such that $r - e$ might be a code word. Even if the decoder produces an error word $e(x)$, we have to further test if the **assumption** of the limited number of errors, the *numerical condition*, is truly satisfied by testing if $e(x)$ is the correct error word by checking $c = r - e$ to see if it is a code word using the *check polynomial*. If it is, then we pass on to the block of *code word*. If it is not, then the decoding fails, and we return an *error* message.

Let us concentrate on the *decoder* part. We assume that there are u erasures and at most v errors. Let

$$e(x) = r(x) - c(x) = e_0 + e_1 x + \cdots + e_{n-1} x^{n-1}$$

be the hypothetical *error vector*. Note that we assume that there are at most $v + u$ non-zero e_i's. It follows from the **remark** after Proposition 1.25 that if we assume that $2u + 2v < \delta$ and $c(\gamma^i) = 0$ for $i = 1, \ldots, \delta$, it means that $e(\gamma^i) = r(\gamma^i)$ for $i = 1, \ldots, \delta$; then, there is a unique code word within the error range. Certainly, by *brute force* of checking all possibilities, we may recover $c(x)$. However, this is too slow. We use the improved *numerical condition $2v + u < \delta$*, which is better than $2u + 2v < \delta$ and find a clever way of solving the decoding problem. We introduce (following Peterson [30]) the following concepts of *error locator* which gives the locations of errors and *error-evaluator polynomial* $\omega(x)$ which gives the values at the error locations.

Definition 3.10. Let M be the set of all places where either there is an erasure or an error. We shall write the set M as the disjoint union $M = N \cup L$, with N consisting of u erasures and L consisting of at most v errors. The *error-locator polynomial* $\sigma(x)$ is defined as

$$\sigma(x) = \prod_{i \in M} (1 - \gamma^i x) = \prod_{i \in N} (1 - \gamma^i x) \prod_{i \in L} (1 - \gamma^i x) = \sigma_1(x)\sigma_2(x).$$

Since the set N is known, then the function $\sigma_1(x)$ is a known polynomial of degree u. Thus, $\sigma(x)$ and $\sigma_2(x)$ determine each other. The *error-evaluator polynomial* $\omega(x)$ is defined as

$$\omega(x) = \sum_{i \in M} (e_i \gamma^i x \prod_{j \in M \setminus i} (1 - \gamma^j x)).$$

Note that $\omega(x)$ is an unknown function. ∎

It is easy to see that $\deg(\sigma_1(x)) = u, \deg(\sigma_2(x)) \leq v, \deg(\sigma(x)) \leq u + v$ and $\deg(\omega(x)) \leq u + v$. Note that if $\sigma_2(x), \omega(x)$ are found, then the solutions of $\sigma_2(x)$ will give us $\{\gamma^{-i} : i \in L\}$. We take the inverses of all elements in the preceding set, and we have $\{\gamma^i : i \in L\}$ and the locations of all errors. Furthermore, we have $-\frac{\omega(\gamma^{-i})\gamma^i}{\sigma'(\gamma^{-i})} = e_i$ for all erasure or error position i. Therefore, if we can find $\sigma_2(x)$ and $\omega(x)$, then we can find $\sigma(x)$ and $\omega(x)$, and we decode the message.

Furthermore, it suffices to find the rational function $\frac{\omega(x)}{\sigma_2(x)}$ with the denominator having a constant 1, instead of finding them individually, for the following two reasons: (1) The only factor of $\sigma_2(x)$ is of the form $(1 - \gamma^i x)$ with $i \in L$ which is not a factor of $\omega(x)$; therefore, the above expression is in the reduced form of the said rational function. (2) The denominator $\sigma_2(x)$ is further normalized by $\sigma_2(0) = 1$. Therefore, once we find the rational function $\frac{\omega(x)}{\sigma_2(x)} \in \mathbf{F}_q[[x]]$, it follows from Proposition 2.36 and the following Proposition 3.11 that if we write it in the reduced form and require it to satisfy condition (2) of Proposition 2.36, then we have the polynomials $\omega(x), \sigma_1(x), \sigma_2(x)$ and solve the decoding problem.

Now we shall apply Proposition 2.36. The degree of $\omega(x) \leq v + u$, and the polynomial $\sigma_1(x)$ is known with degree $s = u$ and the degree of $\sigma_2 \leq v$. We have

$$v + u + v + u - u = 2v + u.$$

Now, we only use the following **numerical condition** for decoding:

$$2v + u = \delta' < \delta.$$

Also, the rational function $\frac{\omega(x)}{\sigma_2(x)}$ is unknown at the very beginning. We have the following proposition which tell us that the first δ' coefficients of the power-series expansion of $\frac{\omega(x)}{\sigma_2(x)}$ satisfy the following equation with known right-hand side. Furthermore, by the result of Proposition 2.36, the rational function $\frac{\omega(x)}{\sigma_2(x)}$ which exists theoretically is thus uniquely determined.

Proposition 3.11. *We have the following expression in* $\mathbf{F}_q[[x]]$:

$$\mathrm{ord}\left(\frac{\omega(x)}{\sigma(x)} - \sum_{j=1}^{\delta'} r(\gamma^j)x^j\right) \geq \delta' + 1.$$

Proof. We have

$$\frac{\omega(x)}{\sigma(x)} = \sum_{i \in M} \frac{e_i \gamma^i x}{1 - \gamma^i x} = \sum_{i \in M} e_i \sum_{j=1}^{\infty} (\gamma^i x)^j$$

$$= \sum_{j=1}^{\infty} \left(\sum_{i \in M} e_i \gamma^{ij}\right) x^j = \sum_{j=1}^{\infty} e(\gamma^j)x^j.$$

Note that $c(\gamma^j) = 0$ for $j = 1, \ldots, \delta'$ and $e(\gamma^j) = r(\gamma^j)$ for $j = 1, \ldots, \delta'$. Therefore, we have

$$\mathrm{ord}\left(\frac{\omega(x)}{\sigma(x)} - \sum_{j=1}^{\delta'} r(\gamma^j)x^j\right) = \mathrm{ord}\left(\sum_{j=1}^{\infty} e(\gamma^j)x^j - \sum_{j=1}^{\delta'} r(\gamma^j)x^j\right)$$

$$\geq \delta' + 1. \qquad \square$$

On the other hand, Proposition 2.36 shows that the rational function $\frac{\omega(x)}{\sigma_2(x)}$ is thus uniquely defined. Sometimes, one may establish the uniqueness proposition first (as we prove Proposition 2.36 first), and then, we tie the object in an equation (as we prove Proposition 2.36). Finally, we use the equation to show the existence. For decoding purpose, we need a fast way to recover the rational function $\frac{\omega(x)}{\sigma_2(x)}$ from Proposition 3.11 (please see the next proposition).

The above equation in the proposition, written slightly differently as follows, is named by Berlekamp as the *key equation* (see Appendix D):

$$(1 + S(x))\sigma(x) \equiv \omega(x) \mod x^{\delta'+1}.$$

Let us postpone the discussion of the method of Berlekamp to Appendix D. Recall that $\sigma_1(x)$ is a known function. Although a piece of partial power-series expansion of $\frac{\omega(x)}{\sigma_2(x)}$ is known by the preceding proposition, we have to recover $\frac{\omega(x)}{\sigma(x)}$ fast. The process is as follows. Let $f_1(x)$ be the polynomial with $\deg(f_1(x)) < \delta' + 1$ and

$$f_1(x) = \sigma_1(x) \sum_{j=1}^{\delta'} r(\gamma^j) x^j \bmod(x^{\delta'+1}). \tag{1^0}$$

Note that since $\sigma_1(x)$ is known, then $f_1(x)$ is known and thus **uniquely determined**. The conclusion of the above proposition can be re-written as

$$\omega(x) = \sigma(x) \left(\sum_{i=1}^{\delta'} r(\gamma^i) x^i \right) + x^{\delta'+1} h(x)$$

$$= \sigma_2(x) f_1(x) + x^{\delta'+1} h^*(x).$$

Therefore, $\omega(x) \in$ the ideal $(f_1(x), x^{\delta'+1}) \subset F_q[x]$. Note that $\sigma_2(0) = 1$, i.e., $\sigma_2(x)$ is an unit in $\mathbf{F}_q[[x]]$. We have the following interesting equation:

$$\frac{\omega(x)}{\sigma_2(x)} = f_1(x) \bmod(x^{\delta'+1}). \tag{1^{00}}$$

In fact, due to the uniqueness result of **Proposition 2.19**, the *long algorithm* applied to $f_1(x)$ and $f_2(x) = x^{\delta'+1}$ will provide a fast way to find the rational function $\frac{\omega(x)}{\sigma_2(x)}$. It turns out to be one of the most useful tools in decoding. We have the following proposition.

Proposition 3.12 (*Euclidean Algorithm with stopping strategy*) (Sugiyama–Kasahara–Hirasawa–Namakawa). *Let us assume that there are non-negative integers u, v and $\delta' = 2v + u$ and polynomials $\omega, \sigma = \sigma_1 \sigma_2$. Let $\deg(\omega(x)) = \delta'$, $\deg(\sigma_1) = u$ and $\deg(\sigma_2) = v$, where $\sigma_1(x)$ is a known function defined by Definition 3.10. We use the notation of* **Proposition 2.19**. *Let $f_1(x)$ be defined above in equation (1^0), and let $f_2(x) = x^{\delta'+1}$, and let $f_i(x)$, n_i be defined as in Proposition 2.16. Let (the stopping time) t be determined as $t = 3$ if $n_3 \le v + u$; otherwise, t is determined by $n_{t-1} > v + u \ge n_t$. For the case $t = 3$, we have*

$$\alpha_3(x) f_1 + \gamma_3(x) f_2(x) = 1 f_1(x) + 0 f_2(x) = f_3(x).$$

Otherwise, (for $t > 3$), in the following equation:

$$\alpha_t(x) f_1(x) + \gamma_t(x) f_2(x) = f_t(x), \tag{1}$$

we have

$$\frac{\omega(x)}{\sigma_2(x)} = \frac{f_t(x)}{\alpha_t(x)}.$$

Furthermore, we have $\sigma_2(x) = h(0)^{-1}\alpha_t(x)$, *where* $\alpha_t(x) = x^s h(x)$ *with* $h(0) \neq 0$ *and* $\omega(x) = h(0)^{-1}f_t(x)$.

Proof. We like to factor out $\alpha_t(x)$ from the above equation (1) and try to get our conclusion directly. However, there are several technicalities. We want to show that the operation of factoring can be performed in the power-series ring $\mathbf{F}_q[[x]]$. Note that we may assume $2v + u = \delta'$, and according to Proposition 2.16, we have $\deg(\alpha_t(x))(\leq \delta' + 1 - n_{t-1}) \leq v=n$ as in **Proposition 2.19** and $\deg f_t(x) = n_t \leq u+v=m$ as in **Proposition 2.36** in any case. Let $\alpha_t(x) = x^s h(x)$ with $h(0) \neq 0$ and $h(x)$ be a unit in $\mathbf{F}_q[[x]]$. Note that then,

$$ord(\alpha_t^{-1}) = -s$$

in $\mathbf{F}_q((x))$, where s may be 0. Then, $s \leq v < 2v + u + 1 = \delta' + 1$; therefore, x^s is a factor of $f_2(x)$, and it follows from the above equation (1) that x^s is a factor of $f_t(x)$. We may factor $\alpha_t(x)$ from equation (1) above and get a power-series expression in $F_q[[x]]$ for the left-hand side of equation (1). We have

$$ord\left(\frac{f_t(x)}{\alpha_t(x)} - f_1(x)\right) = ord(\gamma_t(x)f_2(x)\alpha_t^{-1}) \geq 2v + u + 1 - s$$

$$= (v + u) + v - s + 1.$$

It follows from the above equation (1^{00}) that we have

$$ord\left(\frac{\omega(x)}{\sigma_2(x)} - f_1(x)\right) \geq 2v + u + 1.$$

It follows from **Proposition 2.36, (1)** that

$$\frac{\omega(x)}{\sigma_2(x)} = \frac{f_t(x)}{\alpha_t(x)}.$$

Since $\frac{\omega(x)}{\sigma_2(x)}$ is the reduced form of the rational function, and the condition of $\sigma_2(0) = 1$ makes the rational expression unique, our proposition follows easily. $\qquad\square$

From the above proposition, we easily find $\frac{\omega(x)}{\sigma(x)} = \frac{\omega(x)}{\sigma_1(x)\sigma_2(x)}$ with $\sigma_1(x)$ known.

The following example will help us to understand the procedure.

Remark: The assumptions of the preceding proposition are satisfied if there are less than or equal u erasures and at most v errors. ∎

Example 3: Let us continue our study of the **example** in Section 3.1. It is easy to see that the *generator polynomial*$= g(x) = 1 + x^3 + x^6 + x^9 + x^{12}$. With the number $\delta = 5$, we expect to correct two errors. Let the code word be $c(x) = x + x^2 + x^4 + x^5 + x^7 + x^8 + x^{10} + x^{11} + x^{13} + x^{14}$ and the received word be $r(x) = x + x^2 + x^3 + x^4 + x^5 + x^6 + x^7 + x^8 + x^{10} + x^{11} + x^{13} + x^{14}$. We wish to recover the code word $c(x)$ from the received word $r(x)$. Let us assume that there is no erasure $u = 0$ (i.e., $\sigma_1(x) = 1, \sigma(x) = \sigma_2(x)$), and $v = 2$. We take $\delta' = 4$.

We have the following computations:

$$r(\gamma) = \gamma^2, \quad r(\gamma^2) = \gamma + 1,$$
$$r(\gamma^3) = \gamma, \quad r(\gamma^4) = \gamma^2 + 1.$$

We shall start the *long algorithm* with $(f_1, f_2) = (\gamma^2 x + (\gamma + 1)x^2 + \gamma x^3 + (\gamma^2 + 1)x^4, x^5)$. We have

$$1 \cdot f_1 + 0 \cdot f_2 = 1 \cdot (\gamma^2 x + (\gamma + 1)x^2 + \gamma x^3 + (\gamma^2 + 1)x^4) + 0 \cdot x^5$$
$$= (\gamma^2 x + (\gamma + 1)x^2 + \gamma x^3 + (\gamma^2 + 1)x^4 = f_3,$$
$$f_2 + (1 + \gamma^7 x)f_3 = x^5 + (1 + \gamma^7 x)(\gamma^2 x + (\gamma + 1)x^2 + \gamma x^3 + (\gamma^2 + 1)x^4)$$
$$= \gamma^2 x + \gamma^{14} x^2 + \gamma^6 x^3 = f_4,$$
$$f_3 + \gamma^2 x f_4 = \gamma^2 x = f_5$$
$$(1 + \gamma^2 x + \gamma^2(\gamma^3 + \gamma + 1)x^2)f_1 + \gamma^2 x f_2 = \gamma^2 x = f_5.$$

We conclude that

$$\frac{\omega(x)}{\sigma(x)} = \frac{\gamma^2 x}{1 + \gamma^2 x + (\gamma^2(\gamma^3 + \gamma + 1))x^2}.$$

It means that we have

$$\omega(x) = \gamma^2 x$$
$$\sigma(x) = 1 + \gamma^2 x + (\gamma^2(\gamma^3 + \gamma + 1))x^2.$$

We find the two roots of $\sigma(x)$ are γ^{-3}, γ^{-6} (the simple way is computing all $\sigma(\gamma^{-i})$ which takes at most 15 evaluations).

We have to find e_3, e_6. We calculate $\frac{\omega(x)}{x\sigma'(x)}$ at $x = \gamma^{-3}, \gamma^{-6}$, and both e_3, e_6 are 1. Therefore, we find $c(x) = r(x) - e(x) = x + x^2 + x^4 + x^5 + x^7 + x^8 + x^{10} + x^{11} + x^{13} + x^{14}$ is a possible code word. Further test by multiplying it with the *check polynomial* $= x^3 + 1$ yields 0 mod $x^{15} + 1$ or $c(x) = (x + x^2)(1 + x^3 + x^6 + x^9 + x^{12}) = (x + x^2)g(x)$. We conclude $c(x)$ is a code word. We find the original message is $x + x^2$. ∎

Exercises

(1) Show that a Hamming code is a cyclic code.
(2) Find a BCH code with minimal distance $> \delta$.
(3) Write a computer program to decode the example in Section 3.2.
(4) Find an example of a received word with more than three errors in the example in Section 3.2 which is decoded to a wrong code word.
(5) Find the probability of error that the received word r has more than two errors, for the example in Section 3.2. We assume that the probability of error at any location is p.

3.3. Reed–Solomon Codes

The BCH codes introduced in the preceding sections are general with the disadvantages of being complicated and hard to use. We discuss their simple counterparts, the Reed–Solomon codes, in this section.

The reader is referred to the beginning of Chapter 1. There, we discuss a possible way of coding using polynomial curves over real numbers. The process is as follows. Let $[a_0, a_1, \ldots, a_{k-1}]$ be the original message. It determines a polynomial curve $f(x) = \sum_0^{k-1} a_i x^i$ of degree at most $k - 1$. Then, we send out $[b_1, b_2, \ldots, b_n] = [f(1), f(2), \ldots, f(k), \ldots, f(n)]$. Assuming that there are at most $\lfloor \frac{n-k}{2} \rfloor$ errors, we may use brute force of checking all possibilities to decode the received message $[b_1', b_2', \ldots, b_n']$ as follows. Knowing k, n, we may try all subsets of $[b_1', b_2', \ldots, b_n']$ of $k + \lfloor \frac{n-k}{2} \rfloor$ elements to see which one determines a polynomial curve of degree at most $k - 1$. Once we find the curve, if it exists, we may use it to correct all errors. The difficulty is the decoding process since a brute force way will be time-consuming when n, k are large. Now, we shall change the coefficient field from the real field $\mathbb{R}$ to a finite field $\mathbf{K}$ and rename it the Reed–Solomon code (see the following definition). The important aspect of it is that there is a fast way of decoding (as seen in the following).

Now, we shall use a finite field $\mathbf{F}_q$ in place of $\mathbf{F}_2$, where $q = 2^m$. We shall use the notation $n = q - 1$. Let γ be a primitive element of the field $\mathbf{F}_q$ (i.e., the powers of γ exhaust all non-zero elements of $\mathbf{F}_q$). We shall consider the points $\{\gamma, \gamma^2, \ldots, \gamma^n\}$ instead of $\{1, 2, \ldots, n\}$ in the real case. Note that $\gamma^n = 1$.

We shall define a code. Let $a = [a_0, a_1, \ldots, a_{k-1}]$ be a message. We want to make a code $[b_1, b_2, \ldots, b_n]$. Let $\mathbf{P}_k$ be the set of all polynomials of degree less than k, where $k < n$, over $\mathbf{F}_q$; then, $\mathbf{P}_k$ is a vector space of dimension k over $\mathbf{F}_q$. For any vector $[a_0 a_1 \ldots a_{k-1}] \in \mathbf{F}_q^k$, let us define $f(x) = \sum a_i x^i$, and notice that $f(x) \in \mathbf{P}_k$. Let the values at γ^i be $b_i = f(\gamma^i)$ for $i = 1, \ldots, n$ (note $\gamma^n = 1$). Thus, we define a one-to-one map from the message space $\mathbf{V} = \{a = [a_0 a_1 \ldots a_{k-1}] : a_i \in \mathbf{F_q}\}$ to $\{f(x) = \sum a_i x^i \in \mathbf{P_k}\}$, and then to the code space $\mathbf{C} = \{b = [b_1 b_2 \cdots b_n] : b_i = f(\gamma^i), b \in \mathbf{F_q^n}\}$. We have a code which sends a message $a = [a_0, a_1, \ldots, a_{k-1}]$ to a code $b = [b_1, b_2, \ldots, b_n]$ this way.

Proposition 3.13. *Let us use the notations of the preceding paragraph. The set $\{[b_1 b_2 \ldots b_n] : $ there is a polynomial $f(x) \in \mathbf{P}_k$ with $b_i = f(\gamma^i)\}$ is a k-dimensional subspace $\mathbf{C}$ of the vector space of all n tuples.*

Proof. Note that $deg(f(x)) < k < n$. Let us define a map $\pi: \mathbf{P}_k \to \mathbf{F}_q^n$ as $\pi(f(x)) = [f(\gamma), f(\gamma^2), \ldots, f(\gamma^n)]$. Then clearly, π is a linear transformation. The only thing we have to show is that π is an one-to-one map, i.e., if $f(\gamma^i) = 0$ for $i = 1, 2, \ldots, n$, then $f = 0$. Note that then $x^n - 1 \mid f(x)$, and $deg(f(x)) \geq n$, if $f \neq 0$. A contradiction. $\square$

We have the following definition.

Definition 3.14. The code defined by $\mathbf{C}$ in the preceding proposition is called a **Reed–Solomon code** $[n, k]$ **in the value form**, or simply **Reed–Solomon code** $[n, k]$. ∎

Proposition 3.15. *Let us fix a Reed–Solomon code $[n, k]$ $\mathbf{C}$. Let b, b^* be two distinct vectors in $\mathbf{C}$. Then, $d(b, b^*) \geq n - k + 1$.*

Proof. Let $b = [b_1 b_2 \cdots b_n] \neq b^* = [b_1^* b_2^* \cdots b_n^*]$ and $b_i = f(\gamma^i), b_i^* = f^*(\gamma^i)$. Since $b \neq b^*$, then $f(x) \neq f^*(x)$. Therefore, $(f - f^*)(x)$ has at most degree of $k - 1$ and at most $k - 1$ zeroes or at least $n - k + 1$ non-zeroes among $\{1, \gamma, \gamma^2, \ldots, \gamma^{n-1}\}$. $\square$

It follows from the **remark** after Proposition 1.25 that the Reed–Solomon $[n, k]$ code can correct up to (and including) $\lfloor \frac{n-k}{2} \rfloor$ errors.

Proposition 3.16. *A Reed–Solomon code $[n, k]$ is an $[n, k, n-k+1]$ code.*

Proof. It follows from the preceding proposition that $d \geq n - k + 1$. Let $f(x) = \prod_{i=1}^{i=k-1}(x - \gamma^i)$. Then, it is clear that $f(x) \in P_k(x)$ and $f(\gamma^i) = 0$ for $i = 1, \ldots, k - 1$. Let $b_j = f(\gamma^j)$. We must have $b_j \neq 0$ for $j = k, \ldots, n$; otherwise, its Hamming weight is less than $n - k + 1$. Then, the Hamming weight of $[b_1 \ldots b_n]$ is precisely $n - k + 1$. $\qquad \square$

Proposition 3.17. *A Reed–Solomon code in the value form is an MDS code over $\mathbf{F}_q$.*

Proof. It follows from the definition of *MDS code* of Section 1.7. $\qquad \square$

Remark: Let us assume that $q > 2$, where $q = 2^m$. Although the Reed–Solomon code in the value form is an *MDS code* over the field F_q, it is not an *MDS code* over the bit field F_2. Note that if we consider a Reed–Solomon code in the value form over the bit field, it will be an $[mn, mk]$ code, with the minimal distance $n - k + 1 \leq (n - k)m + 1$. Therefore, it is not an *MDS code* over the bit field $\mathbf{F_2}$ (cf. the definition of *MDS code* in Section 2.4). Recall that according to Shannon's theorem, to get a sequence of good codes, it is unlikely to have any restriction of the sizes of the blocks. Suppose we keep the rate of information $k/n = \ell$ constant, then the rate of distance $((1 - \ell + 1/n)n)/nm$ will go to zero as $m \to \infty$, i.e., the correcting power will tend to zero if we allow the size of the field q to get arbitrarily large. Thus, the Reed–Solomon codes in the value form are not the ones forecasted by Shannon's theorem. An additional drawback is that the multiplications over a large field will become more time-consuming to carry out. $\qquad \blacksquare$

Another way of defining a Reed–Solomon code is to imitate a BCH code as follows. Note that over $\mathbf{F}_q$, all irreducible polynomials of $\{\gamma^j\}$ are of degree 1.

Definition 3.18. A **Reed–Solomon code** $[n, k]$ **in the coefficient form**, or simply **Reed–Solomon***$[n, k]$ **code**, is a primitive BCH code with length $n = q - 1$. The generator polynomial is of the form $g(x) = \prod_{i=1}^{i=d-1}(x - \gamma^i)$, where $d = n - k + 1$. Any message word is a polynomial $h(x)$ of degree less than $k = n - d + 1$, and its code word is the coefficients of the polynomial $c(x) = h(x) \cdot g(x)$. $\qquad \blacksquare$

Example 4: Apparently these two codes, Reed–Solomon code and Reed–Solomon* code, are seemingly different. However, the results may be the same. For instance, let us consider the case $k = 1$ $(d = n)$ the code word for $f(x) = 1$ in a Reed–Solomon code is $[11 \cdots 1]$, and the code word for $h(x) = 1$ in a Reed–Solomon* code $[\alpha_1 \ldots \alpha_n]$, where α_i is given by the coefficients in the expansion of $1 \times \prod_{i=1}^{i=n-1}(x - \gamma^i) = 1 + 1x + 1x^2 + \cdots + 1x^{n-1}$. ∎

We note that the dimension of a Reed–Solomon*$[n, k]$ code is $n - (d - 1) = k =$ dimension of Reed–Solomon code $[n, k]$. In fact, we have the following proposition.

Proposition 3.19. *Let* **C** *be the code space of a Reed–Solomon code* $[n, k]$ *and* **C*** *be the code space of the corresponding Reed–Solomon* * $[n, k]$ *code. Then,* **C** = **C***.

Proof. We wish to show that $\mathbf{C} \subset \mathbf{C}^*$. Since they are of the same dimension, then they must be equal. Let us use linear map: $\mathbf{C} \mapsto \mathbf{C}^*$. Let us consider a basis $\{x^j : 0 \leq j \leq k - 1\}$ in the message space of **C**. The code word of x^j in **C** is $[\gamma^j \gamma^{2j} \ldots \gamma^{nj}]$. Let us interpret it as the coefficients of the polynomial $c(x) = \sum_{i=1}^{n} \gamma^{ji} x^i = \sum_{i=1}^{n} (\gamma^j x)^i$ (Recall that $x^n = 1$). It is easy to see the following identity:

$$c\left(\frac{x}{\gamma^j}\right) = x\frac{x^n - 1}{x - 1} = x \prod_{i=1}^{i=n-1} (x - \gamma^i).$$

Replacing x by $x\gamma^j$ in the above equation, note that $x^n = 1$ and

$$\gamma^{jn} x^n = x^n$$

$$\gamma^j x - 1 = \gamma^j (x - \gamma^{n-j}).$$

Certainly, $n - j \geq n - k + 1 = d$, and recall that the generate polynomial $g(x) = \prod_{i=1}^{i=d-1}(x - \gamma^i)$, we conclude

$$c(x) = c\left(\frac{\gamma^j x}{\gamma^j}\right) = \gamma^j x\frac{x^n - 1}{\gamma^j x - 1} = x \prod_{i \neq n-j} (x - \gamma^i)$$

and

$$c(x) = x \prod_{i \neq (n-j)} (x - \gamma^i) = g(x)h(x).$$

Therefore, $c(x) \in (g(x))$, and the coefficients of $c(x)$ which corresponds to x^j being an element in $\mathbf{C}^*$. Since it is true for any x^j, we conclude that $\mathbf{C} \subset \mathbf{C}^*$. Our proposition is established. $\square$

As illustrated by the preceding proposition, the two different forms of Reed–Solomon codes are equivalent and will be called the Reed–Solomon code. The Reed–Solomon code will later be generalized to a *geometric* Goppa code (a type of *algebraic geometric* code) (see Example 1 in Chapter 5). **The Reed–Solomon code* can be decoded as a BCH code.**

Example 5: Let us consider the following Reed–Solomon* code over $\mathbf{F}_{2^4}$ and $d = 9$. Then, we have $q = 16$, $n = 16 - 1 = 15$, $k = 7$. The generator polynomial $g(x)$ is given by

$$g(x) = \prod_{i=1}^{i=8}(x - \gamma^i),$$

where γ satisfies the equation $x^4 + x + 1 = 0$ and is given in **Example 2** of Section 3.1. According to the theory of Reed–Solomon codes, the code may correct up to 4 errors. So we assume that $v = 4$, and there is no erasure, $u = 0$.

Let us consider a received message $r(x) = 1 + \gamma^6 + \gamma x + \gamma^3 x^2 + \gamma^7 x^3 + \gamma^8 x^4 + \gamma^3 x^5 + x^6 + \gamma^3 x^7 + \gamma x^8 + \gamma^2 x^9 + \gamma^3 x^{10} + \gamma^3 x^{11} + \gamma x^{12} + \gamma^2 x^{13} + \gamma^{14} x^{14}$. We compute $r(\gamma^i)$ as follows:

$$
\begin{aligned}
r(\gamma) &= \gamma^3, \quad r(\gamma^2) = \gamma, \\
r(\gamma^3) &= \gamma^7, \quad r(\gamma^4) = \gamma^7, \\
r(\gamma^5) &= \gamma, \quad r(\gamma^6) = 0, \\
r(\gamma^7) &= \gamma^9, \quad r(\gamma^8) = 0.
\end{aligned}
$$

We start the long algorithm with $f_1 = \sum_{j=1}^{j=8} r(\gamma^j)x^j = \gamma^3 x + \gamma x^2 + \gamma^7 x^3 + \gamma^7 x^4 + \gamma x^5 + \gamma^9 x^7$ and $f_2 = x^9$ as follows:

$$1 \cdot f_1 + 0 \cdot f_2 = f_3,$$

$$f_2 + (\gamma^{13} + \gamma^6 x^2)f_3 = \gamma x + \gamma^{14} x^2 + \gamma^6 x^3 + \gamma^{13} x^4 + \gamma^2 x^5 + \gamma^{13} x^6 = f_4,$$

$$f_3 + (1 + \gamma^{11} x)f_4 = \gamma^9 x + \gamma^2 x^2 + \gamma x^4 + \gamma^6 x^5 = f_5,$$

$$f_4 + (\gamma^9 + \gamma^7 x)f_5 = \gamma^9 x + \gamma^8 x^2 + \gamma^5 x^3 + \gamma^9 x^4 = f_6.$$

Notice that $\deg(f_5) = 5 > v + u = 4 \geq \deg(f_6)$, then we shall stop. We substitute back and have the following equation:

$$((\gamma^9 + \gamma^7 x) + (\gamma^{13} + \gamma^6 x^2) + (\gamma^9 + \gamma^7 x)(1 + \gamma^{12} x)(\gamma^{13} + \gamma^6 x^2))f_1$$

$$+(1 + (\gamma^9 + \gamma^7 x)(1 + \gamma^{11} x))f_2$$

$$= f_6 = \gamma^9 x + \gamma^8 x^2 + \gamma^5 x^3 + \gamma^9 x^4$$

with

$$\sigma(x) = \alpha_u(x) = (\gamma^9 + \gamma^7 x) + (\gamma^{13} + \gamma^6 x^2)$$

$$+ (\gamma^9 + \gamma^7 x)(1 + \gamma^{11} x)(\gamma^{13} + \gamma^6 x^2),$$

$$\omega(x) = f_u(x) = \gamma^9 x + \gamma^8 x^2 + \gamma^5 x^3 + \gamma^9 x^4.$$

It is easy to see that $\sigma(\gamma^{-5}) = \sigma(\gamma^{-7}) = \sigma(\gamma^{-10}) = \sigma(\gamma^{-11}) = 0$. Therefore, the error locations are $5, 7, 10, 11$. Furthermore, we use the formula that $-\frac{\omega(\gamma^{-i})\gamma^i}{\sigma'(\gamma^{-i})} = e_i$ for error locations i. We find that $e_5 = 1$, $e_7 = \gamma^2$, $e_{10} = 1$, $e_{11} = \gamma^2 + 1$. We conclude that the error polynomial $e(x)$ and the code polynomial $c(x)$ are

$$e(x) = x^5 + \gamma^2 x^7 + x^{10} + (\gamma^2 + 1)x^{11},$$

$$c(x) = r(x) - e(x) = 1 + \gamma^6 + \gamma x + \gamma^3 x^2 + \gamma^7 x^3 + \gamma^8 x^4$$

$$+ (\gamma^3 + 1)x^5 + x^6 + (\gamma^2 + \gamma^3)x^7 + \gamma x^8 + \gamma^2 x^9$$

$$+ (1 + \gamma^3)x^{10} + (\gamma^3 + \gamma^2 + 1)x^{11} + \gamma x^{12} + \gamma^2 x^{13} + \gamma^{14} x^{14}.$$

The corrected $c(x)$ is a code polynomial, which can be checked by multiplying with the following check polynomial:

$$\prod_{i=9}^{15}(x - \gamma^i),$$

and show the result is a multiple of $x^{15} + 1$ as follows:

$$((1 + \gamma + \gamma^3) + \gamma^3 x + (\gamma + \gamma^2)x^2 + \gamma^3 x^3$$

$$+ (1 + \gamma^2 + \gamma^3)x^4 + (1 + \gamma^3)x^5 + (1 + \gamma^3)x^6)(x^{15} + 1).$$

Therefore, the original message is

$$(1 + \gamma + \gamma^3) + \gamma^3 x + (\gamma + \gamma^2)x^2 + \gamma^3 x^3 + (1 + \gamma^2 + \gamma^3)x^4$$

$$+ (1 + \gamma^3)x^5 + (1 + \gamma^3)x^6.$$

∎

3.4. Classical Goppa Codes

Instead of using polynomials, we may use rational functions. We study
a class of codes, the classical Goppa[2] codes, which are generalizations of
BCH codes. It will be clear from the following definitions that most *classical
Goppa codes* are not cyclic codes. In Chapter 4, we show that both *classical
Goppa codes* (see Chapter 5, Example 3) and *Reed–Solomon codes* (see
Chapter 5, Example 1) can be generalized to *geometric Goppa codes*.

Let $g(x)$ be any non-constant polynomial $\in \mathbf{F}_q[x]$. We shall consider the
ring $\mathbf{F}_q[x]$. Let $\gamma \in \mathbf{F}_q$ such that $g(\gamma) \neq 0$. By the Euclidean algorithm,
there exist $\alpha(x)$ and $r \neq 0$ such that

$$g(x) = \alpha(x)(x - \gamma) + r.$$

It means that we have $g(\gamma) = r \neq 0$, $\alpha(x) = \frac{g(x) - g(\gamma)}{(x - \gamma)}$, and

$$(-r)^{-1}\alpha(x)(x - \gamma) = 1 \quad \mod \ g(x).$$

Namely, $(x - \gamma)$ has an inverse $(-r)^{-1}\alpha(x) \mod g(x)$. We say that $(x - \gamma)$
is regular in $\mathbf{F}_{q^m}[x]/(g(x))$. It is easy to see that in the polynomial ring
$\mathbf{F}_q[x]$, we have

$$\frac{-1}{g(\gamma)}\left(\frac{g(x) - g(\gamma)}{x - \gamma}\right)(x - \gamma) \equiv 1 \quad \mod \ g(x).$$

Therefore, in the total quotient ring of $\mathbf{F}_{q^m}[x]/(g(x))$, we have

$$\frac{1}{\overline{x - \gamma}} = \frac{-1}{g(\gamma)}\left(\overline{\frac{g(x) - g(\gamma)}{x - \gamma}}\right),$$

or we may write

$$\frac{1}{x - \gamma} \equiv \frac{-1}{g(\gamma)}\left(\frac{g(x) - g(\gamma)}{x - \gamma}\right) \quad \mod \ g(x). \tag{1}$$

We have the following definition.

Definition 3.20. Let $\gamma_i \neq \gamma_j$ be all distinct and $\mathbf{G} = \{\gamma_1, \ldots, \gamma_n\} \subset \mathbf{F}_q$.
We define the *(classical) Goppa code* $\Gamma(\mathbf{G}, g(x))$ with *Goppa polynomial*
$g(x)$, where $g(\gamma_i) \neq 0$ for $1 \leq i \leq n$ to be the set of code words

---

[2]Valery Goppa (1939–), Soviet and Russian mathematician.

$\mathbf{c} = [c_1, \ldots, c_n]$ over the letter field $\mathbf{F}_q$ for which

$$\sum_{i=1}^{n} \frac{c_i}{(x - \gamma_i)} \equiv 0 \mod(g(x)).$$ ∎

The classical Goppa code will later be generalized to a *geometric* Goppa code (a type of *algebraic geometric* code) (see Example 3 in Chapter 5).

The following example will show that the BCH codes are special cases of classical Goppa codes.

Example 6: Let the Goppa polynomial $g(x)$ be x^{d-1}, and $\mathbf{G} = \{\gamma^{-i} : 1 \leq i \leq n-1\}$, where γ is a primitive $(n-1)$th root of unity in $\mathbf{F}_q$. Let us consider the resulting classical Goppa code $\Gamma(\mathbf{G}, g)$. Note that $g(\gamma^{-i}) = (\gamma^{-i})^{d-1}$. Using the equation (1) above, we have the following equations all mod $(g(x))$:

$$\sum_{i=1}^{n-1} \frac{c_i}{(x - \gamma^{-i})} \equiv 0 \Leftrightarrow \sum_{i=1}^{n-1} \frac{(-c_i)}{(\gamma^{-i})^{d-1}} \frac{x^{d-1} - (\gamma^{-i})^{d-1}}{x - \gamma^{-i}} \equiv 0$$

$$\Leftrightarrow \sum_{i=1}^{n-1} \frac{(-c_i)}{(\gamma^{-i})^{d-1}} \sum_{k=0}^{d-2} (\gamma^{-i})^{d-2-k} x^k = 0$$

$$\Leftrightarrow \sum_{i=1}^{n-1} (-c_i) \sum_{k=0}^{d-2} (\gamma^{i(k+1)}) x^k = 0$$

$$\Leftrightarrow \sum_{k=0}^{d-2} \left(\sum_{i=1}^{n-1} (-c_i)(\gamma^{i(k+1)}) \right) x^k = 0.$$

Since it is a polynomial of degree less that $d - 1 = \deg g(x)$, we must have all coefficients zeroes, i.e.,

$$\sum_{i=1}^{n-1} (-c_i) \gamma^{i(k+1)} = 0, \quad \text{for } k = 0, \ldots, n - 2.$$

Let a polynomial $c(x) = \sum_{i=1}^{n-1} (-c_i) x^i$. Then, we have $c(\gamma^{(k+1)}) = 0$ for $k + 1 = 1, \ldots, d - 1$. We conclude that the last condition is precisely the condition for a BCH code. ∎

It is clear that a classical Goppa code is linear. It is important to find other parameters. We have the following proposition.

Proposition 3.21. *Let $g(x)$ be a polynomial of degree s. Then the classical Goppa code $\Gamma(\mathbf{G}, g)$ has rank $= n - s$ (i.e., the rank is the dimension of the code space) and minimum distance $\geq s + 1$.*

Proof. Let us follow the definition for classical Goppa Codes. From the equation

$$\sum_{i=1}^{n} \frac{c_i}{(x - \gamma_i)} \equiv 0 \mod(g(x)),$$

we have that the left-hand side can be written as $n(x)/d(x)$ with $d(x) = \prod(x - \gamma_i)$ and deg $(n(x)) < n$. Since $(d(x), g(x)) = 1$, then $c = \{c_1, \ldots, c_n\}$ is a code word $\Leftrightarrow g(x) \mid n(x)$ or $n(x) = g(x)h(x)$, where $h(x)$ is of degree $< n - s$. Since the classical Goppa codes $\mathbf{C}$ are parameterized by the coefficients of $h(x)$, its rank is $n - s$. On the other hand, if $c = [c_1, \ldots, c_n]$ is a code word with at most s of the c_i's non-zero, then

$$\sum_{i=1}^{n} \frac{c_i}{(x - \gamma_i)} \equiv 0 \mod(g(x))$$

can be rewritten as

$$\sum_{i=1}^{n} \frac{c_i}{(x - \gamma_i)} = \sum_{i \in I} \frac{c_i}{(x - \gamma_i)},$$

where I is an index set with at most s elements. It can be written as $n(x)/d(x)$ with $\deg(n(x)) < s$ and $g(x) \mid n(x)$. Therefore, $n(x) = 0$. Due to the linear-independence property of $\frac{1}{(x - \gamma_i)}$, all $c_i, i \in I$ must be zero. Hence, we conclude that the minimum distance $\geq s + 1$. $\qquad \square$

The next proposition shows that the classical Goppa codes satisfy the Gilbert–Varshamov's bound.

Proposition 3.22. *For any n, let $\mathbf{G}_n = \mathbf{F}_{q^n}$. Then, there is a sequence of classical Goppa codes $\Gamma(\mathbf{G}_n, g_n(x))$ for some monic irreducible polynomials $g_n(x)$ such that $\lim_{n \mapsto \infty} R(\Gamma(\mathbf{G}_n, g_n(x))) \geq 1 - H(\delta)$, where $R(M)$ is the rate of information of the code M and δ is any number with $0 \leq \delta \leq \frac{q-1}{q}$. It means that the sequence of classical Goppa codes meets the Gilbert–Varshamov's bound.*

Proof. Given δ with $0 \leq \delta \leq \frac{q-1}{q}$, let $d_n < n$ be a sequence of integers such that

$$\lim_{n \mapsto \infty} \frac{d_n}{n} = \delta.$$

Given any positive integer m, let t_n be a sequence of integers such that $t_n \leq \frac{n}{m}$. We shall find an irreducible polynomial $g_n(x)$ of degree t_n such that the minimal Hamming distance of $\Gamma(\mathbf{G}_n, g_n(x))$ is at least d_n. For this purpose, all code words with Hamming weight less than d_n should be excluded from $\Gamma(\mathbf{G}_n, g_n(x))$. Let $(c_1, \ldots, c_n)$ be such a code word, i.e.,

$$\sum_{i=1}^{n} \frac{c_1}{(x - \gamma_i)} \equiv 0 \mod (g_n(x)).$$

We may write the above rational function as $\frac{c(x)}{h(x)}$, and the denominator $h(x)$ is co-prime to $g_n(x)$, which means $g_n(x) \mid c(x)$. Let us consider a code word $[c_1, \ldots, c_n]$ of Hamming weight j; it means that $\deg c(x) \leq (j - 1)$, and thus, $c(x)$ has at most $\lfloor \frac{j-1}{t_n} \rfloor$ irreducible factors of degree t_n. The total number of code words of Hamming weight j is $\mathbf{C}_j^n (q - 1)^j$. Therefore, the total number of irreducible polynomials which should be excluded is, by Proposition 2.30,

$$\sum_{j=1}^{d_n - 1} \left\lfloor \frac{j-1}{t_n} \right\rfloor \mathbf{C}_j^n (q - 1)^j < \frac{d_n}{t_n} \mathbf{U}_q(n, d_n - 1).$$

Let us count the total number of irreducible polynomials of degree t_n. By Proposition 2.30, with t_n the composite number,

$$\mathbf{I}_{t_n} > \frac{1}{t_n} q^{mt_n} (1 - q^{-mt_n/2+1}).$$

To prove our proposition, it suffices to show that asymptotically, we may find d_n, t_n with

$$\frac{d_n}{t_n} \mathbf{U}_q(n, d_n - 1) < \frac{1}{t_n} q^{mt_n} (1 - q^{-mt_n/2+1}).$$

Let us take a logarithm with base q and divide by n. Furthermore, let $m \mapsto \infty$, $(n \mapsto \infty)$ and $\frac{d_n}{n} \mapsto \delta$ (note that then $\frac{d_n - 1}{n} \mapsto \delta$). It follows from Proposition 2.30 that we are looking for

$$H_q(\delta) \leq \lim_{n \mapsto \infty} \frac{mt_n}{n}.$$

We may select t_n to make the above an equality. Therefore, we show the existence of the required irreducible polynomial $g_n(x)$. Further, note that it follows from **Proposition 3.21** that the information rate, $R(\Gamma(\mathbf{G}_n, g_n(x)))$, of the classical Goppa code $\Gamma(\mathbf{G}_n, g_n(x))$ is at least $1 - \frac{mt_n}{n}$, and we prove that

$$\lim_{n \mapsto \infty} R(\Gamma(\mathbf{G}_n, g_n(x))) \geq 1 - H_q(\delta). \qquad \square$$

3.4.1. *Decoding Classical Goppa Codes*

Our process of decoding classical Goppa codes is similar to the decoding process of a BCH code (see Section 3.2). Let $g(x)$ be a polynomial of degree $2t$. Then, the classical Goppa code $\Gamma(\mathbf{G}, g)$ has a minimal distance $\geq 2t+1$. We expect to correct t errors. The process of decoding is as follows.

Let $r = [r_1, \ldots, r_n]$ be the received word. Define the *syndrome*, $S_r(x)$, as

$$S_r(x) = \sum_{i=1}^{n} \frac{r_i}{x - \gamma_i} \mod(g(x)).$$

If the *syndrome* $S_r(x) = 0$, then r is a code word. If the *syndrome* $S_r(x) \neq 0$, then we apply the following process to decode. Note that $(x - \gamma_i)$ and $g(x)$ are co-primes. Therefore, it follows from equation (1) before Definition 3.20 that

$$\frac{1}{x - \gamma_i} \equiv h_i(x) \mod(g(x)),$$

where $h_i(x)$ is a polynomial of degree less than $2t$. Then, the *syndrome* $S_r(x)$ can be expressed as a polynomial $h(x)$ which is equivalent to $S_r(x)$ mod $(g(x))$ and of degree less than $2t$. The important fact is that $S_r(x)$, and hence $h(x)$, is computable and known. Furthermore, let $r = c+e$ where c is a code word and e is the error word, $M = \{i : e_i \neq 0\} =$ the error locations of r with $cardinality(M) \leq t$. Then, $S_c(x) = 0$ and

$$S_r(x) = S_e(x) = \sum_{i \in M} \frac{r_i}{x - \gamma_i} = \sum_{i \in M} r_i h_i(x) \mod(g(x)) = h(x) \mod g(x).$$

As in the decoding a BCH code, let $\alpha_r(x) = \prod_{i \in M}(x - \gamma_i)$ be the error locator of r and $\omega_r(x) = \sum_{i \in M} e_i \prod_{j \in M \setminus i} (x - \gamma_j)$ with degree $< t$ be the error evaluator. Then, we have the following *key equation*:

$$S_r(x)\alpha_r(x) = \sum_{i \in M} \frac{r_i}{x - \gamma_i} \prod_{i \in M}(x - \gamma_i) = \sum_{i \in M} \frac{e_i}{x - \gamma_i} \prod_{i \in M}(x - \gamma_i)$$
$$= \alpha_r(x)h(x) \mod(g(x)) = \omega_r(x) \mod(g(x)).$$

The decoding is to find $\alpha_r(x)$ and $\omega_r(x)$. Since $S_r(x)$ is equivalent to $h(x) \mod(g(x))$, we conclude that $\omega_r(x)$ is in the ideal generated by $h(x)$ and $g(x)$. We use long algorithm to find $\omega_r(x)$ and $\alpha_r(x)$ in a routine way as follows. Let $f_1(x) = h(x)$ and $f_2(x) = g(x)$. Now, note that

$\deg(h(x)) < 2t = \deg(g(x))$. We have

$$1 \cdot f_1(x) + 0 \cdot f_2 = f_3$$

$$\cdots\cdots$$

$$\alpha_u f_1(x) + \gamma_u f_2(x) = f_u(x)$$

$$\cdots\cdots,$$

where u is the first time that $n_u = \deg(f_u(x)) < t$, i.e., $n_{u-1} = \deg(f_{u-1}(x)) \geq t$. We have by Proposition 2.19 that $\deg(\alpha_u(x))(\leq 2t - n_{u-1}) \leq t$. Combining the above equations, we have

$$\alpha_r(x)(\alpha_u(x)f_1(x)) = \alpha_u(x)\omega_r(x) = \alpha_r(x)f_u(x) \quad \mod \ (g(x)).$$

Since the multiples of the polynomials $\alpha_u(x)\omega_r(x)$ and $\alpha_r(x)f_u(x)$ are of degrees $< 2t = \deg(g(x))$, we conclude

$$\alpha_u(x)\omega_r(x) = \alpha_r(x)f_u(x),$$

$$\frac{f_u(x)}{\alpha_u(x)} = \frac{\omega_r(x)}{\alpha_r(x)}.$$

Now, it is a simple matter to finish the decoding process in the following way: (1) Throw away the G.C.D. of $\alpha_u(x)$ and $f_u(x)$, and call the resulting polynomials $\alpha_u^*(x), f_u^*(x)$. Furthermore, we make $\alpha_u^*(x)$ monic and adjust $f_u^*(x)$ without changing the fraction. We shall still call them $\alpha_u^*(x), f_u^*(x)$, then we have $\alpha_r(x) = \alpha_u^*(x)$ and $\omega_r(x) = f_u^*(x)$. (2) Find all roots of $\alpha_r(x)$, we have M. Let γ_i be a root of $\alpha_r(x)$. Compute $\omega_r(\gamma_i)$. Then, we have $e_i \prod_{j \in M \setminus i}(\gamma_i - \gamma_j)$ and hence e_i. We correct the word by taking $c = r - e$ and check c to see if it is a code word. If it is, then we are done. If not, then there are more than t errors, and we return an *error* message.

Exercises

(1) Decode Reed–Solomon $[15, 5, 11]$ code.

(2) Decode the classical Goppa code in the example of Section 3.4.

(3) Write a computer program to decode Reed–Solomon $[15,7,9]$ code.

(4) We have a document to be preserved. The document is of length $70,000$ letters (including all blanks, punctuation, etc.). If the decaying rate is 1% for every ten years, and if in case that 30% of the letters are ruined, then the whole document is unreadable. How long will the document last if it is written plainly or if it is written with Reed–Solomon code $[15,7,9]$?

PART III

Algebraic Geometry

Chapter 4

Algebraic Geometry

In the last chapter, we discussed the *Reed–Solomon codes*, which are used to evaluate a polynomial of degree at most $k - 1$ at $n(\geq k)$ points in the ring of polynomials $\mathbf{F}_q[x]$, where $\mathbf{F}_q$ is a field with q elements. We see that it is equivalent to evaluate $L(\mathbf{D})$ on $P^1_{\mathbf{F}_q}$ (see Example 1 of Section 5.1) with $\mathbf{D} = nP_\infty$ (see Section 4.3). Similarly, *classical Goppa code* can be considered as a code over $P^1_{\mathbf{F}_q}$. We may extend the concept of *Reed–Solomon codes* and *classical Goppa codes* to codes over any projective smooth curve (instead of lines only). The Riemann–Roch theorem induces a richer algebraic structure, and the corresponding codes will be more useful.

Now we advance to the next level of coding theory, in which we focus on *geometric Goppa codes*. Our attention on the rings $\mathbf{F}_q[x]$ will be refocused on *algebraic functions of one variable*, i.e, function rings of curves. We need some knowledge of algebraic geometry, a rich and beautiful subject. It has been there for two thousand years. Great minds of the past have rebuilt it again and again. Some guiding lights from the past will be treated as simple corollaries of theorems, and the foundations of algebraic geometry might come last.

We will not try to give a comprehensive description of algebraic geometry in this book. Instead, we emphasize the useful curve theory, especially *Riemann–Roch theorem* and Weil's theorem on the *zeta function*, which gives us a count on the number of *rational points* on a smooth curve over a finite field $\mathbf{F}_q$, and simply discuss it to the extent that the reader will be able to bear the burden with us. Sometimes for the readers' understanding of the subject, we have to give the general picture of algebraic geometry which may not be relevant to the study of coding theory. Most proofs will be deferred to the standard books at hands such as *Commutative Algebra* [16] by Zariski and Samuel, *Introduction to the Theory of Algebraic Functions of One Variable* [10] by Chevalley, *Introduction to Algebraic Geometry* [13] by Mumford, and *Algebraic Geometry* [26] by Harshorne. We give some examples to illustrate the points.

Usually we consider an *algebraically closed* ground field in algebraic geometry, but in the context of coding theory, we have to discuss the case that the ground field is *finite*, and hence, we have to consider a field which is **not** algebraically closed. In this survey chapter, we assume that the ground field is either *algebraically closed* (which includes the complex field $\mathbb{C}$) or the real field $\mathbb{R}$ or *finite field* $\mathbf{F}_q$ if not stated explicitly otherwise.

4.1. Affine Spaces and Projective Spaces

In **plane Euclidean geometry** with real axis, we discuss quadratic curves: ellipses, parabolas, hyperbolas. That is the beginning of algebraic geometry. In the general setup which we intend to discuss, we: (1) replace the *real number* axis by an axis over any *field*, (2) replace the words *plane* (i.e., two-dimensional) by any finite-dimensional *affine* or *projective* space, (3) replace a single quadratic equation by any system of polynomial equations (of any finite degrees).

The point (1) above is natural. The point (2) of increasing the dimension is natural too. Let us consider the extension to *projective spaces*. The usual point set $\{[a_1, \ldots, a_n] : a_i \in \mathbf{K}\}$ (without the usual vector space structure) is called an n-dimensional *affine space* $A^n(\mathbf{K})$ over the field $\mathbf{K}$.

It is well known that two distinct parallel lines in an affine plane will never meet or only meet at infinity. If we add the points at infinities, then all pairs of two lines will meet, at affine plane or the infinities. Following the notations of Zariski and Samuel [16] (Vol. II, p. 127), we have the algebro-geometric concept of *completeness*, which is unrelated to the topological concept of completeness (which is defined by using Cauchy sequences which in general, for an abstract field $\mathbf{K}$, does not exist). In algebraic geometry, we define that an algebraic set is said to be *complete* if every valuation of the quotient field has a center in the set. For instance, an affine space is **not complete** because the valuation defined by the subspace at infinity has no center at the affine space. [1]

---

[1] Also see Munford [13] Section 9, Chapter 1.

The set of all points in affine space $A^n(\mathbf{K})$ plus the points at infinity is called n-dimensional *projective space* $P^n(\mathbf{K})$. It can be show that *projective space* $P^n(\mathbf{K})$ is *complete*. Rigorously, we define an equivalence relation $\approx$ in the $(n+1)$-dimensional affine space $A^{n+1}(\mathbf{K})\backslash\{0\}= \{(a_0, a_1, \ldots, a_n) :$ *not all* $a_i = 0\}$ as follows:

$$(a_0, a_1, \ldots, a_n) \approx (b_0, b_1, \ldots, b_n) \Leftrightarrow a_i = tb_i \quad \text{for some common } t \neq 0.$$

We define the n-dimensional *projective space* $P^n(\mathbf{K})$ as $(A^{n+1}(\mathbf{K})\backslash\{0\})/\approx$. There is a natural embedding of $A^n(\mathbf{K})$ into $P^n(\mathbf{K})$ by sending $(a_1, \ldots, a_n)$ to $(1, a_1, \ldots, a_n)$. Let us further explain this point of view.

It is easy to see that $P^0(\mathbf{K})$ is just the point $\{a_0 : a_0 \neq 0\}/\approx$. For $P^1(\mathbf{K})$, all points can be grouped into two sets: $\{(1, a_1) : a_1 \in \mathbf{K}\} \cup (\{(0, a_1) : a_1 \neq 0\}/\approx \ \{(1, a_1) : a_1 \in \mathbf{K}\} \cup P^0(\mathbf{K}) = A^1(\mathbf{K}) \cup P^0(\mathbf{K})$. Let us consider the following picture.

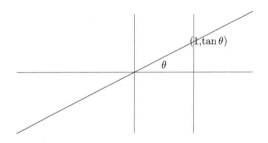

Let us consider the real field $\mathbb{R}$ case. The projective line $P^1(\mathbb{R})$ can be represented by all lines in the plane passing through the origin. Let the line and the horizontal line span an angle θ. As long as $\theta \neq 90°$, the line will have an intersection $(1, \tan\theta)$ with the vertical line $(1, t)$; therefore, all lines passing through $(0,0)$ can be represented as the vertical line $x = 1$ union with one extra point $P^0(\mathbf{K})$ or the point at ∞ corresponding to $\theta = 90°$. Therefore, $A^1(\mathbb{R})$ is the real line, and $P^1(R)$ is a cycle. If the ground field is the complex field $\mathbb{C}$, then $A^1(\mathbb{C})$ is the complex line which is a real plane, and $P^1(\mathbb{C})$ is the one-point compactification of the real plane, i.e., a sphere. In general, for any field $\mathbf{K}$, we have

$$P^n(\mathbf{K}) = A^n(\mathbf{K}) \cup P^{n-1}(\mathbf{K}).$$

In the case the ground field is the real field $\mathbb{R}$, we have $P^2(\mathbb{R}) = A^2(\mathbb{R}) \cup P^1(\mathbb{R})$, i.e., an affine plane with a circle attached at infinity. We have the following picture for $P^2(\mathbb{R})$.

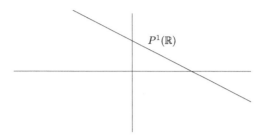

The above does not enlighten any non-expert. Topologically, we may view $P^2(\mathbb{R})$ as a unit closed disc (the open disc is homeomorphic to $A^2(\mathbb{R})$) with the antipodes boundary points identified (which is homeomorphic to a circle, note that two antipodes determine a line). The interior of the disc is identified with $A^2(\mathbb{R})$ by $v \mapsto v/(1 - |v|)$, where v is an vector inside the unit disc and $|v|$ is the length of v. The real project plane is a non-orientable surface.

From the point of view of algebraic geometry, the affine spaces are not *complete* [2] (in the sense of valuation) over any field $\mathbf{K}$, while the projective spaces are.

We define a *linear subspace* $\mathbf{L}$ in $P_{\mathbf{K}}^n$ as the image of $\mathbf{U} \backslash \{0\}$, where $\mathbf{U}$ is a positive-dimensional subspace of the $A_{\mathbf{K}}^{n+1}$ treated as a vector space. We define $\dim(\mathbf{L}) = \dim(\mathbf{U}) - 1$.

We have the following well-known theorem about vector spaces:

Proposition 4.1. *Let* $\mathbf{U_1}, \mathbf{U_2}$ *be two subspaces of a vector space* $\mathbf{K}^{n+1}$. *Then, we have*

$$\dim(\mathbf{U_1} \cap \mathbf{U_2}) \geq \dim(\mathbf{U_1}) + \dim(\mathbf{U_2}) - (n+1).$$ ∎

Proposition 4.2. *Let* $\mathbf{L_1}, \mathbf{L_2}$ *be two linear spaces of* $P_{\mathbf{K}}^n$. *Then, we have*

$$\dim(\mathbf{L_1} \cap \mathbf{L_2}) \geq \dim(\mathbf{L_1}) + \dim(\mathbf{L_2}) - n.$$ ∎

Corollary 4.3. *Let* $\mathbf{L_1}, \mathbf{L_2}$ *be two linear spaces of* $P_{\mathbf{K}}^2$ *and* $\dim(\mathbf{L_1}) = \dim(\mathbf{L_2}) = 1$. *Then, the intersection of* $\mathbf{L_1}, \mathbf{L_2}$ *is not empty.*   ∎

Even over a finite field, both $A_{\mathbf{K}}^n$ and $P_{\mathbf{K}}^n$ are finite. However, the distinction pointed out by the preceding Corollary is obvious. We say that

---

[2] In a notation of Zariski and Samuel [16].

projective spaces are *complete*.[3] The above corollary will be generalized to Bézout's Theorem (see Proposition 4.20).

4.2. Affine Algebraic Varieties

Although we only need curve theory for coding theory, we shall enjoy the beautiful theory of *algebraic geometry*. Let us speak further about $A_{\mathbf{K}}^n$. There are two objects involved as follows: (1) *the algebraic objects* of the affine space $A_{\mathbf{K}}^n$, including a polynomial ring $\mathbf{K}[x_1, \ldots, x_n]$ (over the projective space $P_{\mathbf{K}}^n$; there is a *homogeneous* polynomial ring $\mathbf{K}[x_0, x_1, \ldots, x_n]^h$ ={all homogeneous polynomials}), and the system of equations there. (2) *Geometric objects* of the affine space $A_{\mathbf{K}}^n$, which is the solution sets $X(\mathbf{I})$ of ideals $\mathbf{I}$ of equations. Similarly, we have projective geometric objects, i.e., the solution sets $X(\overline{\mathbf{I}})$ of homogeneous ideals of equations. The relation between these two objects may be complicated in general. Primarily, we take the point of view of algebra. We shall study *affine varieties* first.

We have the following well-known Hilbert basis theorem.

Proposition 4.4 (Hilbert basis theorem). *The ideal generated by any system of polynomials in n variables $\mathbf{K}[x_1, \ldots, x_n]$ can be generated by finitely many polynomials.*

Proof. Zariski and Samuel Vol. I [16]. □

Note that $X(\mathbf{I})$ may be empty even in the simplest case that $\mathbf{I} = (f)$, where $f(x_1, \ldots, x_n) \neq$ non-zero constant. Look at the following examples.

Example 1: In general there is no one-to-one relation between the algebraic objects and the geometric objects. Let $\mathbf{K} = \mathbb{R}$ the real field and $f(x_1, \ldots, x_n) = x_1^2 + 1$, then there is no solution. Similarly, the equation $1 = 0$ has no solution, and it is easy to see that the ideals $(f) \neq (1)$. Therefore, the different ideals may produce the same solution sets. However, if we extend the field $\mathbb{R}$ to the complex field $\mathbb{C}$, then there are two solutions $x_1 = i, -i$, and the equation $1 = 0$ still has no solution, and sometimes we cannot tell the ideals apart by looking at the solution sets. ∎

Example 2: It is very important to know the number of solutions of equations for the construction of *algebraic-geometric coding*. This is a non-trivial problem. We want to indicate that sometimes there is no solution

---

[3]In the sense of Zariski and Samuel [16].

for a non-trivial equation. Let $\mathbf{F}_q$ be a finite field with q elements, where $q = p^m$, and let $x_0^{q-1} + \cdots + x_n^{q-1} = 0$, where $n+1$ is less than p and greater than 1 be a homogeneous polynomial, Since $x_i^{q-1} = 1$ for any $x_i \neq 0$. Then, in $P_\mathbf{K}^n$, there is no solution. However, over a suitable finite field extension of $\mathbf{F}_q$, there will be solutions.   ∎

We want to use algebra to study geometric objects. It would be better to establish an one-to-one correspondence between the algebraic objects and the geometric objects. Then, we have to go to some suitable extension of the ground field $\mathbf{K}$. If we consider all algebraic objects as rings, ideals etc., it might be simpler to consider the ultimate algebraic field extension, the *algebraic closure* of $\mathbf{K}$. In algebraic geometry, we usually assume that the field $\mathbf{K}$ is algebraically closed from the very beginning. However, in coding theory, we only consider the finite fields for the letter fields; hence, we may not consider an algebraically closed field only. Furthermore, if we only consider a fixed ideal $\mathbf{I}$, we may be satisfied with some finite field extension of $\mathbf{K}$. The advantage is that if $\mathbf{K}$ is a finite field, then a finite extension of it stays finite. Therefore, we may stay with finite and non-algebraically closed fields. There are many properties of finite fields we can use.

Even we consider finite field sometimes, and we may study the cases that the ground fields are algebraically closed. Let us see what happens if we replace the field $\mathbf{K}$ by its algebraic closure $\bar{\mathbf{K}}$ or simply assume that $\mathbf{K}$ is algebraically closed. Then, we have the following well-known theorem. We need the following preliminary definition.

Definition 4.5. Let $\mathbf{I}$ be an ideal of a given ring R, the radical of $\mathbf{I}$, Rad $\mathbf{I}$, is defined to be

$$\mathbf{I} \subset Rad\ \mathbf{I} = \{f : f^m \in \mathbf{I}, \text{for some positive integer } m\}.$$

We define a *radical ideal* $\mathbf{I}$ to be an ideal $\mathbf{I}$ which equals $Rad\ \mathbf{I}$.   ∎

We have the following important proposition.

Proposition 4.6 (Hilbert's nullstellensatz). *Let $\mathbf{K}$ be an algebraically closed field. Let $\mathbf{I}$ be an ideal of $\mathbf{K}[x_1, \ldots, x_n]$ and S any subset of $A_\mathbf{K}^n$. Let*
$X(\mathbf{I}) = \{(a_1, \ldots, a_n) : (a_1, \ldots, a_n) \in \mathbf{K}^n, g(a_1, \ldots, a_n) = 0,\ for\ all\ g \in \mathbf{I}\}$ *and*
$\mathbf{J}(S) = \{g : g \in \mathbf{K}[x_1, \ldots, x_n], g(a_1, \ldots, a_n) = 0,\ for\ all\ (a_1, \ldots, a_n) \in S\}.$
Then,

$$Rad\ \mathbf{I} = J(X(\mathbf{I})) = \{f \in \mathbf{K}[x_1, \ldots, x_n] : f(a_1, \ldots, a_n) = 0,$$

$$for\ all\ (a_1, \ldots, a_n) \in S\}.$$

Proof. Zariski and Samuel Vol. II, p. 164 [16]. $\qquad\square$

Example 3: Let $\mathbb{C}$ be the field of complex numbers. Note that $\mathbb{C}$ is algebraically closed. Let $\mathbf{I}$ be a proper ideal (i.e., $\mathbf{I} \neq (1)$). Then, $\mathbf{I} \subset$ some maximal ideal $\mathbf{M}$. Let $\mathbf{I} \subset \text{Rad } (\mathbf{I}) = \mathbf{I}' \subset \mathbf{M}$. We have $\mathbb{C}[x_1, \ldots, x_n]/\mathbf{M} \equiv \mathbb{C}$. Let $(\bar{x}_1, \ldots, \bar{x}_n) = (a_1, \ldots, a_n)$. Then clearly, $\mathbf{M} = (x_1 - a_1, \ldots, x_n - a_n)$, and $f(a_1, \ldots, a_n) = 0$ for all $f \in \mathbf{I}'$. It means that the ideal $\mathbf{I}'$ (hence $\mathbf{I}$) has at least one common solution. $\qquad\blacksquare$

If we have an algebraically closed ground field $\mathbf{K}$, then we define an *affine algebraic variety* to be the solution set of a radical ideal in $\mathbf{K}[x_1, \ldots, x_n]$. Then, the above proposition implies that in the case of algebraically closed ground fields $\mathbf{K}$, there is a natural one-to-one correspondence between the set of algebraic varieties ($=$ the sets of solutions of system of equations from radical ideals) and the set of radical ideals. It means that in this setup, the *algebraic objects* and the *geometric objects* reflect each other.

Exercises

(1) We define JacRad($\mathbf{I}$)$= \cap_{\mathbf{m} \in X(\mathbf{I})} \mathbf{m}$. Any ring with the property that $JacRad(\mathbf{I}) = Rad(\mathbf{I})$ for all ideals $\mathbf{I}$ will be called a *Hilbert* ring (or a *Jacobson* ring). Show that the power series ring $\mathbf{K}[[x]]$ is not a *Hilbert's* ring.

(2) Let $f = x^2 y^3 \in \mathbf{K}[x, y]$. What is $Rad\ (f)$?

(3) Show that $Rad((x^2 + 1)) \neq Rad((1))$ in the ring $\mathbb{R}[x]$, where $\mathbb{R}$ is the field of real numbers.

(4) Find a finite set of generators for the ideal generated by $\{x^i - y^{i+1} : i = 10, 11, \ldots\}$ in the polynomial ring $\mathbf{K}[x, y]$ of two variables.

(5) Is the ideal $(x^1 + 1, y^2 + 1) \in \mathbb{R}[x, y]$ prime? radical? where $\mathbb{R}$ is the field of real numbers.

4.3. Regular Functions and Rational Functions

From *complex analysis* on the extended complex numbers $(\mathbb{C} \cup \{\infty\}) = \overline{\mathbb{C}}$ (or $P_1(\mathbb{C})$ which is a projective space)), we have three kinds of functions to be considered:

(1) All meromorphic functions without poles. By Liouville's theorem, these consist of all constant functions.

(2) All meromorphic functions with finitely many poles. It can be shown that this is the set of all rational functions $\mathbb{C}(x)$ in one variable.

(3) All meromorphic functions with infinitely many poles, i.e., we include all meromorphic functions with essential singularities. The Picard's great theorem told us that possibly except one value, the function will take any value infinitely many times. It is not well studied.

Apparently the first set is too narrow; the field of constants is common to many projective curves and will not tell the the underlining geometric sets apart. It may not be useful for our study. The third set is too big and not well studied. Riemann selected the second set and proved an important theorem (see Proposition 4.28) which tells the underlining geometric sets apart by the concept of *genus* (see Proposition 4.28). Since then, the field of rational functions become an indispensable tool of algebraic geometry.

In the present book of *algebraic coding theory*, we need the results of *Riemann–Roch* theorem over algebraic functions of one variable over a finite field. This was started in the work in 1882 by Dedekind[4] and Weber.[5] Even the concept of divisors were theirs. The form of *Riemann–Roch theorem* we need is due to Weil.[6]

If we consider all rational functions of any algebraic curve C over a field **K**, then we have an infinite-dimensional vector space. We want to classify them. We define two algebraic curves to be birationally isomorphic iff their rational function fields are isomorphic over **K**.

Definition 4.7. Two curves C, C' are said to be *birationally equivalent* iff the rational function fields $\mathbf{F}(C), \mathbf{F}(C')$ of them are **K**-isomorphic.   ∎

However, it is only a *qualitative statement* about birational isomorphisms. Riemann ingeniously pick up rational functions with no worse poles than the prescribed ones. He was able to show that it is a finite-dimensional vector space, and the dimension is tied to a global topological invariant, the *genus*. Later on, we establish the following *Riemann's inequality*, where $d(\mathbf{D})$ is the *degree* of a *divisor* $\mathbf{D}$ (see Definition 4.22) and $\ell(\mathbf{D})$ is the dimension of the $L(\mathbf{D})$ (see Definition 4.22):

$$\ell(\mathbf{D}) \geq d(\mathbf{D}) + 1 - g,$$

---

[4]German mathematician, 1831–1916.
[5]German mathematician, 1842–1913.
[6]André Weil (1906–1998), an influential French mathematician.

where g is a non-negative integer which is the *genus*. Furthermore, if $d(\mathbf{D}) \geq 2g - 1$, then

$$\ell(\mathbf{D}) = d(\mathbf{D}) + 1 - g.$$

One of the central approaches of mathematics is to define a set of functions $R(S)$ on a set S and to study the relations between $R(S)$ and S. We may have *differentiable* or *analytic* functions on a set S, then we may have differentiable or analytic geometry. In algebraic geometry, following Riemann, we are more interested in rational functions which may have *poles* inside a projective variety.

Before we study the *projective varieties* and the fields of rational functions associated with them, let us consider an affine piece, an *affine variety*. Given an affine algebraic variety X defined by an ideal $\mathbf{J}(X)$, we define the *ring of regular functions* $R[X]$ as $\mathbf{K}[x_1, \ldots, x_n]/\mathbf{J}(X) = \mathbf{K}[\bar{x}_1, \ldots, \bar{x}_n]$, i.e., we only consider *induced polynomials* from the affine space $A_{\mathbf{K}}^n$ to X. Naturally, two induced polynomials are considered to be identical iff they have identical values on the algebraic variety, i.e., it is the view that the values $f(\mathbf{P})$ of a function determine a function f.

Let us write $f(\mathbf{P})$ as $[f, \mathbf{P}]$. Over the complex numbers, a function $[f, *]$ is determined by $f(\mathbf{P})$ for all points $\mathbf{P}$. Apré Grothedick,[7] in general, we believe that a point $\mathbf{P}$ is determined by $[*, \mathbf{P}]$ for all functions $* \in R(X)$. Pushing one step further, we may identify a point $\mathbf{P}$ as an evaluation map $\mathbf{P} : R[x] \mapsto R[X]/\mathbf{P}$ a field. Then clearly, we require that $\mathbf{P}$ is a maximal ideal. We may generalize the above considerations to a ring R. We have the following definition.

Definition 4.8. A point P of a given affine algebraic variety X over a field $\mathbf{K}$ with the ring of regular functions $R[X]$ is a maximal ideal $\mathbf{P}$ of $R[X]$. If $\mathbf{K} \equiv R[X]/\mathbf{P}$ under the map, then the point is called a *rational point*. Otherwise, it is a non-rational point. In general, given any noetherian ring R, we may define the points set as $\mathbf{m}\text{-}spec(R) = \{\mathbf{m} : \mathbf{m}$ *maximal ideal of* $R\}$.

Let $X(\mathbf{J}$ be defined as $\{\mathbf{m} : \mathbf{m} \in m\text{-}spec(R)\}$. Let $f \in R$. We define $f(\mathbf{P})$ as the image of f under the canonical map $\pi : R \mapsto R/\mathbf{P}$. ∎

Example 4: Let us consider the ring $\mathbb{Z}$ of all integers. Let $\mathbf{I} = (0)$. Then clearly, all maximal ideals are of the form (p), where p is a prime number.

---

[7]A. Grothendick (1928–2014), stateless-French mathematician, founder of modern Algebraic Geometry.

We have $X((0)) = \{(p) : p\,prime\}$. And $X((p)) = (p)$, $X((12)) = \{(2),(3)\}$. Moreover, we have $Rad((12)) = (2) \cap (3) = (6)$. ∎

Example 5: Let X be the affine line A_R^1 over real field $\mathbb{R}$. Then, $(x^2 + 1)$ is a maximal ideal of $\mathbb{R}[x]$. It is a point. However, $x^2 + 1 = 0$ has no real solution, and it can be shown that $\mathbb{R}[x]/(x^2 + 1) \equiv \mathbb{C} \neq \mathbb{R}$, where $\mathbb{C}$ is the field of complex numbers. So, the point $(x^2 + 1)$ is not a *rational point*. ∎

Definition 4.9. Let $\mathbf{P}$ be a maximal ideal. The algebraic degree of the field extension $[R[X]/\mathbf{P} : \mathbf{K}]$ is called the *residue degree* $\mu(\mathbf{P})$. Rational points are points with residue degree 1. ∎

We have the following proposition.

Proposition 4.10. *A point $\mathbf{P}$ is rational $\Leftrightarrow \mathbf{P} = (x_1 - a_1, \ldots, x_n - a_n)$, where $a_i \in \mathbf{K}$ for all i.*

Proof. ($\Rightarrow$) If $\mathbf{P}$ is rational, then $R[X]/\mathbf{P} = \mathbf{K}$, so $\bar{x}_i = a_i \in \mathbf{K}$ for all i. Furthermore, $(x_1 - a_1, \ldots, x_n - a_n)$ is a maximal ideal, and $\mathbf{P} = (x_1 - a_1, \ldots, x_n - a_n)$.
($\Leftarrow$) If $\mathbf{P} = (x_1 - a_1, \ldots, x_n - a_n)$, then $R[X]/\mathbf{P} = \mathbf{K}$. □

We consider the case that $\mathbf{K}[x_1, \ldots, x_n]/\mathbf{J}(X)$ is an *integral domain*, i.e., $\mathbf{J}(X)$ is a *prime ideal*. Then, the algebraic variety X will be called an *irreducible algebraic variety*. We restrict our attention to only the set of irreducible algebraic varieties, and by abusing the language, using the term *algebraic variety* only for an *irreducible algebraic variety*, if not stated explicitly otherwise. In this case, the *total quotient ring* of $\mathbf{R}[X]$ is a field. We call it the *rational function field* $\mathbf{F}(X)$ of the algebraic variety X. If the ideal $\mathbf{J}(X)$ stays prime for all field extensions of $\mathbf{K}$, then we say the algebraic variety X is *absolutely irreducible*. The ideal $(x^2 + 1)$ of $\mathbb{R}[x]$ is not absolutely irreducible, while the ideal $(x + 1)$ is absolutely irreducible.

Definition 4.11. We define the *local ring* of an affine algebraic variety X at a point $\mathbf{P}$ as $\mathbf{K}[\bar{x}_1, \ldots, \bar{x}_n]_p = \{f/g : f \in \mathbf{K}[\bar{x}_1, \ldots, \bar{x}_n], g \notin \mathbf{P}\}$. Algebraically, we define a *local ring*, $(\mathbf{R}, \mathbf{q})$, as a ring $\mathbf{R}$ with a unique maximal ideal $\mathbf{q}$. It is easy to see that the local ring $(\mathbf{R}, \mathbf{q})$ of an affine algebraic variety X at a point $\mathbf{P}$ is a local ring in the sense of algebra. We recall the definition of its $\mathbf{q}$-adic completion $\bar{\mathbf{R}}$ as the completion of $\mathbf{R}$ with respect to the metric induced by the ideal $\mathbf{q}$ as follows. In general, let

$(\mathbf{R}, \mathbf{q})$ be a local ring; we define the order distance function as

$$\mathrm{ord}_{\mathbf{q}}(f) = \max\{m : f \in \mathbf{q}^m\} \text{ if } f \in \mathbf{F}(X)\backslash 0, \text{ otherwise } \infty.$$

Furthermore, we define the distance between $f, g \in \mathbf{R}$ as

$$d(f, g) = 2^{-\mathrm{ord}_q(f-g)}.$$

It is routine to check that $\mathbf{R}$ forms a metric space with respect to the preceding defined distance function. We *complete* $\mathbf{R}$ as a metric space and call $\bar{\mathbf{R}}$ the $\mathbf{q}$-adic completion. ∎

There is a *dimension* theory of local rings involving the maximal length of chains of prime ideals. The reader is advised to consult Zariski and Samuel [16] for the definition.

Definition 4.12. A local ring $(\mathbf{R}, \mathbf{q})$ of *dimension* m is said to be a *regular* local ring iff its maximal ideal q can be generated by m elements $\{x_1, \ldots, x_m\}$, which are called *uniformization parameters*. In the equicharacteristic case (let the residue field $\mathbf{R}/\mathbf{q}$ be $\mathbf{K}$ and the quotient field of $\mathbf{R}$ and $\mathbf{K}$ have the same characteristic (cf. Zariski and Samuel, Vol. II, p. 304 [16], *Cohen's theorem*), the completion $\bar{\mathbf{R}}$ of $\mathbf{R}$ is a formal power series ring in m variables. In this case, $\bar{\mathbf{R}} = \bar{\mathbf{K}}[[x_1, \ldots, x_m]]$, where $\bar{\mathbf{K}}$ is algebraic over $\mathbf{K}$, and the point is *rational* iff $\bar{\mathbf{K}} = \mathbf{K}$. A point $\mathbf{P}$ is said to be a regular point of an affine algebraic variety X if the *local ring* at $\mathbf{P}$ is regular. ∎

Example 6: Let p be a prime number. Then, the local ring Z_p is regular with quotient field $\mathbb{Q}$ of characteristic 0, while its residue field is of characteristic $p > 0$. Therefore, it is *not* the equicharacteristic case. Its completion is the p-adic numbers which is not a formal power series ring. ∎

Example 7: Let $\mathbf{P} = (x^2 + 1)$ be a maximal ideal in $\mathbb{R}[x]$, where $\mathbb{R}$ is the real number field. Then, the local ring of $\mathbb{R}[x]$ at $\mathbf{P}$ is regular with the residue field $\mathbb{R}[x]/(x^2 + 1) \equiv \mathbb{C}$, where $\mathbb{C}$ is the field of complex numbers. In the completion of $\mathbb{R}[x]_{(x^2+1)}$, let $t = x^2 + 1$, then there is an element

$$\frac{x}{\sqrt{1-t}} = i,$$

the completion of $\mathbb{R}[x]_{(x^2+1)}$ is $\mathbb{C}[[t]]$. ∎

Exercises

(1) Let us consider a curve C over a field $\mathbf{K}$ defined by

$$xy + x^3 + y^3 = 0.$$

Is the local ring of C at $(0,0)$ regular?

(2) Let $\mathbb{R}$ be the field of real numbers. Show that there are no rational points for two varieties X_1, X_2 defined by ideals $(x^2 + 1)$ and (1). Furthermore, show that there are non-rational points in X_1 but not in X_2.

(3) Prove Eisenstein's criterion: Let $f(x,y) \in \mathbf{K}[x,y]$ be written as

$$f(x,y) = \sum_{i=0}^{n} f_i(x)y^i.$$

Suppose that $(f_0(x), f_1(x), \ldots, f_n(x)) = (1)$ and that there exists $b \in L$ which is a finite extension of $\mathbf{K}$ such that

$$f_n(b) \neq 0,$$
$$f_i(b) = 0, \quad for\ i < n,$$
$$(x - b)^2 \nmid f_0(x).$$

Then, $f(x,y)$ is absolutely irreducible.

(4) Using the above Eisenstein's criterion, show that $y^n - x$ is absolutely irreducible.

(5) Let $\mathbf{P}_1$ be the point $(1,0)$ on $P_{\mathbf{K}}^1$ with the local ring $O_{\mathbf{P}_1}$. Let $\mathbf{P}_2$ be the point $(0,1)$ on the curve defined by $x^3 + y^3 = 1$ with the local ring $O'_{\mathbf{P}_2}$. Show that $O_{\mathbf{P}_1}$ is non-isomorphic to $O'_{\mathbf{P}_2}$. Show that $\hat{O}_{\mathbf{P}_1}$ is isomorphic to $\hat{O}'_{\mathbf{P}_2}$.

(6) Let $\mathbf{P} = (x^2 + 1)$ be a maximal ideal in $\mathbf{R}[x]$, where $\mathbb{R}$ is the real number field. Show that the local ring of $\mathbb{R}[x]$ at $\mathbf{P}$ is regular.

4.4. Affine Algebraic Curves

In coding theory, at present, the only part of *algebraic geometry* we use is *algebraic curve theory*. An *algebraic curve* is determined by its rational function field which is of transcendental degree 1 over a ground field $\mathbf{K}$.

We shall start with *affine algebraic geometry*. We present a broad discussion which will help us understand the subjects better and open up possible new unknown applications. We have the following proposition.

Proposition 4.13 (Noether's normalization lemma). *Let R be an integral domain which is a finitely generated extension ring of a field $\mathbf{K}$. Let r be the transcendental degree of the quotient field F of R over $\mathbf{K}$. Then, there is an algebraically independent subset $\{t_1, \ldots, t_r\}$ such that R is integral over $\mathbf{K}[t_1, \ldots, t_r]$.*

Proof. Zariski and Samuel Vol. I, p. 104 [16]. □

Dimension theory is an interesting subject. For linear algebra, we define the *dimension* of a vector space as the number of elements of a basis. In topology, some times, we use the *Cĕch* theory of dimension. In analysis, we have Hausdorff dimension. In algebraic geometry, we define the *dimension* of $A_{\mathbf{K}}^N$ (or $P_{\mathbf{K}}^n$) to be n. The above lemma lays the ground for us to define the *dimension* of an (irreducible affine) algebraic variety X to be the *transcendental degree* of the field $F(X)$ over the field $\mathbf{K}$. A *curve is an algebraic variety X of dimension 1.*

We have the following definition of smoothness by the Jacobian criterion, the readers are referred to Section 2.6 for the definition of derivative.

Definition 4.14. A rational point $x = (a_1, \ldots, a_N)$ is a smooth point of an affine variety $X = X(\mathbf{K}[x_1, \ldots, x_N]/(f_1, \ldots, f_r)))$ of dimension n iff

$$rank[b_{ij}] = N - n,$$

where

$$b_{ij} = \frac{\partial f_i}{\partial x_j}(a_1, \ldots, a_N).$$

Otherwise, it is called a *singular* point.

An algebraic variety is *smooth* if all rational points are smooth points. ■

Proposition 4.15. *For our case of a perfect ground field, a point is regular iff it is smooth.*

Proof. Mumford [13] p. 343. □

Proposition 4.16. *Let $f(x,y) = 0$ define an affine plane curve C. Then, the collection of all the singular points of C, the singular locus, is defined by the ideal $(f(x,y), f_x, f_y)$.*

Proof. It follows from the definition.   □

Proposition 4.17. *Let $f(x,y) = 0$ define an absolutely irreducible affine plane curve C. Then, the affine plane curve C is smooth iff $(f(x,y), f_x, f_y) = (1)$.*

Proof. It follows from the previous proposition.   □

Example 8: Let n be a positive integer such that $p{\nmid}n$. Let us consider the following Fermat curves over $\mathbf{F}_{p^m}$ with equation,

$$x^n + y^n - 1 = 0.$$

It follows from the above proposition that it is smooth for this affine part.

∎

In general, we may discuss even *arithmetic* cases. See the following example.

Example 9: We may consider the plane curve C defined by $f(x,y) = y^2 - x(x-1)(x-a)$ over a field $\mathbf{K}$ of characteristic not 2, where $a \neq 0, 1$. It is easy to see that $(f, f_x, f_y) = (1)$. Hence, the affine curve C is non-singular.

Let us homogenize the equation by introducing a variable z as $g(x,y,z) = y^2 z - x(x-z)(x-az)$. Clearly, $g(x,y,1) = f(x,y)$. It is clear that the affine curves defined by $g(1,y,z)$ or $g(x,1,z)$ are smooth. We say the projective curve defined by $g(x,y,z)$ is smooth.   ∎

Let $\mathbf{R}$ be the rational function field of an algebraic variety over a ground field $\mathbf{K}$. The algebraic closure of $\mathbf{K}$ in the quotient field of $\mathbf{R}$ is called the *field of constants*. For simplicity, we assume that $\mathbf{K}$ = the field of constants, i.e., every element in $\mathbf{R}$ outside $\mathbf{K}$ is transcendental over $\mathbf{K}$.

Proposition 4.18. *Let $\mathbf{F}$ be a finitely generated field extension of a field $\mathbf{K}$. If the ground field $\mathbf{K}$ is perfect, then there is a transcendental basis $\{x_1, \ldots, x_n\}$ of $\mathbf{F}$ over $\mathbf{K}$ such that $\mathbf{F}$ is separable over $\mathbf{K}(x_1, \ldots, x_n)$.*

Proof. Zariski and Samuel Vol. I, p. 105 [16].   □

In case that the transcendental degree $n = 1$, since every finite separable extension is a simple extension, if the field $\mathbf{K}$ is perfect, we may reduce the field to $\mathbf{K}(x,y)$. It means that we find a plane model for the curve. Although any curve has a plane curve as a model (i.e., those two curves share the same rational function field, or those two curves are *birationally isomorphic*).

Note that a projective plane curve will have at most $\frac{q^3-1}{q-1} = q^2 + q + 1$ rational points ($P^2_{\mathbf{K}}$ has so many points). A non-plane curve may have many more rational smooth points (cf. Proposition 4.10) . If the number of smooth points matters in our discussion (see the next chapter on *geometric Goppa codes*), then we may not be able to discuss only smooth curves in the plane, while it is known that any curve can be represented as a smooth space curve.

Let C be a curve; recall that the curve is said to irreducible if the ideal $J(C)$ is prime. Let us consider the following example.

Example 10: Let the field $\mathbf{K} = \mathbb{R}$ the field of real numbers. Let $J(C) = (x^2+1)$ in $\mathbb{R}[x,y]$. Then, $J(C)$ is irreducible. However, if we extend the field $\mathbb{R}$ to the complex field $\mathbb{C}$, then the generating polynomial $(x^2 + 1)$ splits into a product $(x+i)(x-i)$. So, the curve splits into two lines. ∎

Definition 4.19. Let the field $\mathbf{K}$ be algebraically closed. Let C be an irreducible plane curves defined by equations $F = 0$ and $\mathbf{P} = (x-a, y-b)$ be a point of $A^2_{\mathbf{K}}$. Let us transform the point to $(0,0)$. Let $f(x,y) = f_m(x,y) +$ *higher terms* with m order of $f(x,y)$ at $(0,0)$. We have the following: (1) C is smooth at $(0,0)$ iff $m = 1$; (2) if $m \geq 2$ and $f_m(x,y)$ splits to **distinct** linear polynomials, then C has *ordinary singularity*. ∎

Example 11: Let $\mathbf{K} = \mathbb{C}$ the complex field. Let F be the following equations:

$$(1)\ F = x^2 + y^2 + x^3,$$
$$(2)\ F = x^2 + y^3.$$

Then, it is easy to see that in case (1), C has an ordinary singularity, and in case (2), C does not have an ordinary singularity, or we say the singularity at $(0,0)$ is not ordinary. ∎

In the next proposition, we show that every curve over an algebraically closed ground field $\mathbf{K}$, any curve C is birationally isomorphic to a plane curve with only ordinary singularities. For a proof, the reader is referred to a book on algebraic curve theory by Walker [15].

Proposition 4.20. *Let the ground field* **K** *be algebraically closed. Then, any curve C over* **K** *is birationally isomorphic to an plane curve which is non-singular or is singular with only ordinary singularities.* ∎

Remark: the above proposition will be used in 4.30 (Plücker's formula) to find the *genus* (see Proposition 4.28) of an algebraic curve. Furthermore, if we discuss the notions of *quadratic transformation* and *monoid transformation* (or *blow-ups*) at those singular points, then it is easy to see that any algebraic curve over an algebraic closed field has a non-singular model.

Example 12: Let **K**= $\mathbb{C}$, $f = x^2 + y^3$. Then, the equation can be rewritten as $(x/y)^2 + y = z^2 + y$ with $x/y = z$, and the field **K**$(\bar{x}, \bar{y})$ can be rewritten as **K**$((\bar{z}), \bar{y})$ = **K**$(\bar{z})$. Therefore, the curve c is birationally isomorphic to a line which is non-singular. ∎

Exercises

(1) Let C be a curve defined over **K** with ch(**K**) $= 3$ by the following equation:
$$xy + x^3 + y^3 = 0.$$
Find the local rings at $(0,0)$ and $(1,1)$ and their completions.

(2) Find the pole set of x for the curve $P_{\mathbf{K}}^1$ with regular function ring **K**$[x]$.

(3) Show that the curve C with defining equation $F = x^n + y^{n+1}$ is birationally isomorphic to a line over any algebraically closed ground field **K**.

(4) Show that $x + y^n$ is absolutely irreducible.

(5) Show that the local rings of $x^3 + y^3 = 1$ at $(1,0)$ and the line x at (0) are not isomorphic while their completions are.

4.5. Projective Algebraic Curves

The algebraic geometry used in coding theory is about smooth projective curves. Let us discuss the historical developments of algebraic geometry. The affine algebraic geometry is incomplete (in the sense of valuations). Desargues[8] created *projective geometry*. Following Poncelet,[9] we defined a

---

[8]G. Desargues (1591–1661), French mathematician.
[9]J. Poncelet (1788–1867), French mathematician.

projective algebraic variety to be an affine algebraic variety with points at infinity added. To be precise, with an ideal $\mathbf{I}$ in $\mathbf{K}[x_1, \ldots, x_n]$, we define the homogeneous ideal $\mathbf{I}^h$ associated with it as follows:

$$\mathbf{I}^h = \{x_0^d f(x_1/x_0, \ldots, x_n/x_0) : f \in \mathbf{I}, d = \deg(f)\},$$

and we define the projective completion as $X(\mathbf{I}^h)$. Then, we define a *projective variety* as the projective completion of an affine algebraic variety. Note that inside the projective variety, the affine piece defined by $x_0 = 1$ is the original affine variety. Similarly, we may take one of $x_1, \ldots, x_n$ to be 1, then we get the other n affine parts of the projective variety. These $n + 1$ affine pieces form an affine covering of the projective variety. A projective variety V is said to be *smooth* iff all its $n + 1$ affine pieces are smooth.

Example 13: Let n be a positive integer such that $p \nmid n$. Let us consider the following *projective* Fermat curves over $\mathbf{F}_{p^m}$ with equation,

$$x_1^n + x_2^n = x_0^n.$$

Let us consider the affine part defined by $x_0 = 1$. Let $x = x_1/x_0$, $y = x_2/x_0$. Then, the above can be rewritten as

$$x^n + y^n = 1.$$

From the discussion of **Example 8** in Section 4.4, we know it is smooth for this affine part. Similarly, for other affine parts, it is smooth. So, it is smooth as a projective curve. ∎

We define the rational functions of an algebraic variety (affine or projective) as elements in the rational function field $\mathbf{F}(X)$. Note that there is a major difference between the concepts of *functions* as in the rational function field and *functions* of calculus (say $C[0, 1]$, the set of continuous functions defined over $[0, 1]$).

Usually, in the second case, functions have a common domain of definition.

The common domain of definition of all rational functions is the empty set. For the rational function field $\mathbf{F}(X)$, there are functions which have a pole at any given rational point $\mathbf{P}$; say, let f be a non-constant rational function not having a pole at $\mathbf{P}$, $f(\mathbf{P}) = a$, then $1/(f(x) - a)$ will have a pole at $\mathbf{P}$. Therefore, the common domain of definition is the empty set, not the algebraic variety X. This is a point which is taken up by *sheaf theory* and makes algebraic geometry more interesting.

Proposition 4.21. *Let the ground field* **K** *be algebraically closed. Then, any curve* C *over* **K** *is birationally isomorphic to an plane curve which is non-singular or is singular with only ordinary singularities.* ∎

Remark: The above proposition will be used in 4.30 (Plücker's formula) to find the *genus* (see Proposition 4.28) of an algebraic curve. Furthermore, if we discuss the notions of *quadratic transformation* and *monoid transformation* (or *blow-ups*) at those singular points, then it is easy to see that any algebraic curve over an algebraic closed field has a non-singular model.

Example 12: Let **K**= $\mathbb{C}$, $f = x^2 + y^3$. Then, the equation can be rewritten as $(x/y)^2 + y = z^2 + y$ with $x/y = z$, and the field **K**$(\bar{x}, \bar{y})$ can be rewritten as **K**$(\overline{(z)}, \bar{y}) = $ **K**$(\bar{z})$. Therefore, the curve c is birationally isomorphic to a line which is non-singular. ∎

We wish to generalize Corollary 4.3 on the intersection of two projective lines on $P_{\mathbf{K}}^2$. Let us consider any two homogeneous forms F, G of degrees m, n, respectively. They define two not necessarily irreducible projective plane algebraic curves C, D of degree m, n, respectively. We say that they have a component if F, G have a non-constant common factor. The intersection multiplicities of the two curves at various intersection points are complicate. Say, a straight line usually intersects a smooth curve C at point **P** with multiplicity 1. However, if the straight line is tangential to the curve at the point **P**, then the intersection multiplicity is at least 2. This is the intersection multiplicity. There is a proper way of counting. The following Bézout's theorem is historically important; however, it will not be used in this book, and we simply list it here. Let us quote the Bézout's theorem as follows.

Proposition 4.22 (Bézout's theorem). *Let* C, D *be two not necessarily irreducible projective plane algebraic curves of degree* m, n, *respectively. If they do not have a common component, then they intersect at* mn *points with intersection numbers properly counted.*

Proof. This theorem follows from Walker's book, *Algebraic Curves* [15]. ∎

Let us see the following example.

Example 14: Let us consider two projective curves $z^{n-1}x + y^n$ and x defined over a field $\mathbf{F}_q$. Then, the intersection multiplicity of these two curves at **P** $= (0, 0, 1)$ is n. ∎

[**Valuations and Places**] We take the point of view of *algebra* in the study of curve theory. We follow the book [10]. First, instead of geometric object, we are given a function field $\mathbf{F}$ of one variable over a perfect field $\mathbf{K}$, i.e., $\mathbf{F}$ contains an element x which is transcendental over $\mathbf{K}$ and $\mathbf{F}$ is algebraic of finite degree over $\mathbf{K}(x)$. Furthermore, we assume that $\mathbf{K}$ is algebraically closed in $\mathbf{F}$.

Zariski[10] (cf. Vol. II, p. 110 [16]) defined the *Riemann surface* of $\mathbf{F}$ as the collection of all $\mathbf{K}$-valuations of $\mathbf{F}$ (see following definition) since all valuations of curves are non-singular local rings (see following for a definition). This shows that the Riemann surfaces are non-singular models of curves. Over the complex field $\mathbb{C}$, the Riemann surface in the sense of Zariski equals to the *Riemann surface* in analysis as sets; however, they are equipped with different topologies.

Let us recall that the $\mathbf{K}$-valuations are defined as the subrings O with the following properties:

(1) $O \supset \mathbf{K}$,
(2) $O \neq \mathbf{F}$, and
(3) if $x \in \mathbf{F} \backslash O$, then $x^{-1} \in O$.

By a *place* $\mathbf{P}$ in $\mathbf{F}$, we mean a subset $\mathbf{P}$ of $\mathbf{F}$ which is the ideal of non-units of some valuation ring O.

[**Divisors**] In fact, all valuations discussed in this book are discrete valuation of rank 1 (cf. [16] Vol. II, p. 42). Let $f \in \mathbf{F} \backslash \mathbf{K}$ be an element and $\mathbf{P}$ a place. Define $\mu_{\mathbf{P}}(f) = \max\{m : f \in \mathbf{P}^m\}$, and we write the *divisor* $(f) = \sum_{\mathbf{P}} \mu_{\mathbf{P}}(f)\mathbf{P}$. Then, we know that the sum is a finite sum. Let us write the *zero divisor* $(f)_0 = \sum_{\mathbf{P}} \mu_{\mathbf{P}}(f)\mathbf{P}$, where $\mu_{\mathbf{P}}(f)$ are non-negative, and $(f)_\infty = -\sum_{\mathbf{P}} \mu_{\mathbf{P}}(f)\mathbf{P}$, where $\mu_{\mathbf{P}}(f)$ are negative. Then clearly, $(f)_0 = (f^{-1})_\infty$ and $(f)_\infty = (f^{-1})_0$. We have the following proposition.

Proposition 4.23. *Let* $f \in \mathbf{F} \backslash \mathbf{K}$. *Let* $(f)_0 = \sum_{\mathbf{P}} \mu_{\mathbf{P}}(f)\mathbf{P}$. *Then,* $\sum_{\mathbf{P}} \mu_{\mathbf{P}}(f) \leq [\mathbf{F} : k(f)]$. ∎

Proof. cf. [10] p. 15, Theorem 4. □

---

[10]Oscar Zariski (1899–1986), Russian-born American mathematician, one of the most influential algebraic geometers of the 20th century.

From the preceding proposition, we conclude that $(f) = (f)_0 - (f)_\infty$ is a divisor in the next definition.

Definition 4.24. A *divisor* is $\mathbf{D} = \sum_i n_i \mathbf{P}_i$, where the sum is a finite formal sum and all n_i are integers and $\mathbf{P}_i$ are places. It is clear that *divisors* form a group $\mathbf{D}(\mathbf{F})$ with respect to addition. If all $n_i \geq 0$, we say that $\mathbf{D}$ is *effective*, in symbol, $\mathbf{D} \succ 0$. We define $L(\mathbf{D})$ as

$$L(\mathbf{D}) = \{ f \in \mathbf{F}(X) : (f) + \mathbf{D} \succ 0 \},$$

and we define $\ell(\mathbf{D}) = \dim_{\mathbf{K}}(L(\mathbf{D}))$, with the degree of $\mathbf{D}$, $d(\mathbf{D}) = \sum n_j [O_j/\mathbf{P}_j : \mathbf{K}] = \sum_j n_j \mu_j$, where O_i is the valuation of $\mathbf{P}_i$, and $\mu_j = [O_j/\mathbf{P}_j : \mathbf{K}]$. ∎

In general, we define the *divisor group*[11] of a smooth projective algebraic curve C as the abelian group generated by all points $\mathbf{P}_i$ (rational or not). The concept of divisor group was courtesy of Dedekind and Weber, who used it to extend the work of Riemann on the complex field to arbitrary field even over a finite field. We define the following.

Example 15: Let us consider the real field $\mathbb{R}$ and the projective line $P^1_{\mathbb{R}}$ over $\mathbb{R}$. Let us consider the function $f = x^2 + 1$. It vanishes at the maximal ideal $\mathbf{P} = (x^2 + 1)$ *w*th residue field degree $\mu_p(f) = 2$, and a double pole at ∞, $\mathbf{P}_\infty$. Therefore, it is divisor $D = (f) = \mathbf{P} - 2\mathbf{P}_\infty$ and degree $d((f)) = 1 \cdot 2 - 2 \cdot 1 = 0$. ∎

We have the following proposition (cf. Chevalley [10], p. 18, Theorem 5).

Proposition 4.25. *The divisor of any element $f \neq 0$ of a field of algebraic functions of one variable is of degree 0.* ∎

Furthermore, we have the following proposition which follows from Chevalley [10], p. 18, Theorem 3 and its corollary. It is a generalization of the classical *Liouville's theorem*.

Proposition 4.26. *Let us assume that the field $\mathbf{K}$ is perfect. Then, every non-constant function f of a project curve has some poles.*

Proof. See Chevalley [10], p. 18, Theorem 3 and its corollary. □

---

[11] Chevalley [10] uses multiplicative format for the divisor group rather than the additive group format for the divisor group in this book. Note that his $\mathcal{L}(\mathbf{D})$ is our $L(-\mathbf{D})$.

Definition 4.27. Let $(f) = \sum n_i \mathbf{P}_i$. The collection of all zeroes and poles will be called the support of f, $\mathrm{supp}(f) = \{\mathbf{P}_i : n_i \neq 0\}$. ∎

Example 16: Let us consider $P_{\mathbb{R}}^1$, where $\mathbb{R}$ is the real field. Let $\mathbf{P} = (x^2 + 1)$, P_∞ be the point at ∞ and $f = x^2 + 1$. Then, $(f)_0 = P$ and $(f)_\infty = 2P_\infty$, while $\mu(P) = 2$. Therefore, $d((f)_0) = 2 = d((f)_\infty)$. ∎

It is easy to see that all non-zero elements f, g in $\mathbf{F}$ form a group, and we have $(fg) = (f) + (g)$. Therefore, $(\{(f) : f \neq 0 \in \mathbf{F}\})$ define a subgroup of the divisor group $\mathbf{D}(\mathbf{F})$ which is an abelian group. Their quotient is called the Picard group of $\mathbf{F}$, i.e., two divisors $\mathbf{D}, \mathbf{E}$ are said to be linearly equivalent iff $\mathbf{D} - \mathbf{E} = (f)$ for some $f \in \mathbf{F}$ and $f \neq 0$. For projective line $\mathbf{P}^1$, since for any two points $\mathbf{P}_1$ and $\mathbf{P}_2$, there is a function f with $(f) = \mathbf{P}_1 - \mathbf{P}_2$. It is easy to see that any two divisors of the same degree are linearly equivalent, and it follows from Proposition 4.24 that two linearly equivalent divisors must be of same degree. Therefore, the *Picard group* of $\mathbf{P}^1$ is isomorphic to the group of integers $\mathbb{Z}$. This is an interesting subject while not useful in coding theory. We need more tools, and will not go further on this subject. In the following example we show that not any two curves are birationally equivalent directly.

Example 17: Let $\mathbf{K}$ be an algebraically closed field field with $\mathrm{ch}(\mathbf{K})=p$, where p may be zero. Let us consider the following Fermat curves F_n:

$$x_1^n + x_2^n = x_0^n,$$

where $n > 2$ and $p \nmid n$ (in case that $p = 0$, the second condition $p \nmid n$ is void).

We claim that F_n is not *birationally equivalent* to P_K^1, or the rational function field $F(F_n) \neq \mathbf{K}(t)$. Suppose it is, we may dehomogenize the defining equation by setting $x_0 = 1$, and let $x_1 = g(t)/f(t), x_2 = h(t)/f(t)$ with $(f(t), g(t), h(t)) = (1)$. We wish to deduce a contradiction if f, g, h are not all in the field $\mathbf{K}$. Suppose that there are triples f, g, h, we take a triple f, g, h with $\max(\deg(f), \deg(g), \deg(h))$ the smallest possible one. Note that if two of f, g, h are constants, then the third one must be constant which is impossible. The above defining equation becomes

$$g(t)^n + h(t)^n = \prod_{i=0}^{n-1} (g(t) + \omega^i h(t)) = f(t)^n,$$

where ω is a nth root of unity. Since each prime factor $p(t)$ of two of $(g(t) + \omega^i h(t))$'s must be a prime factor of g, h and hence a prime factor of f, all $(g(t) + \omega^i h(t))$ are co-prime. We may rewrite the above equation as

$$g(t) + \omega^i h(t) = c_i \alpha_i^n(t)$$

and $\sum \deg(\alpha_i) = \deg(f(t))$. Now, we select three of them with one of α_i non-constant. Take the three corresponding equations and eliminate g, h; we get a new equation of the following form:

$$e_i \alpha_i^n(t) + e_j \alpha_j^n(t) + e_k \alpha_k^n(t) = 0.$$

Absorbing the coefficients e_i, e_j, e_k to the polynomials, we rewrite the above as

$$\beta_i^n(t) + \beta_j^n(t) = \beta_k^n(t).$$

Note that $\max(\deg(f), \deg(g), \deg(h)) > \max(\deg(\alpha_i), \deg(\alpha_j), \deg(\alpha_k))$. A contradiction.   ∎

Exercises

(1) Let us assume that ch($\mathbf{K}$)$>$2. Let us consider a curve, a projective *elliptic curve* $\mathbf{C}$ in the plane, whose affine part is defined by $x_2^2 = (x_1 - 2)(x_1 - 1)(x_1 - 0)$. Show that $\mathbf{C}$ is not birationally equivalent to P_K^1.

(2) Let us consider P_K^1 with the ring of regular functions $\mathbf{K}[x]$. Find the divisor (x).

(3) Let a curve C with its affine part defined by $x^2 + y^3 + x^3 = 0$. What is the equation which defines its projective completion?

(4) Let a projective curve C with its affine part defined by $x^2 + y^3 + x^3 = 0$. What is the divisor (x)?

(5) Find the intersection number $I(\mathbf{P}, F \cap G)$ for $\mathbf{P} = (9, 0)$, $F = x + y + x^3$, $G = x^2 - y^4 + x^3$.

4.6. Riemann's Theorem

An important mathematical approach to the study of geometric sets is to study the relations between the geometric sets and the algebraic objects defined over the sets. One example is the *Poincaré conjecture*: The geometric sets, spheres, and the algebraic objects, the homotopy groups of the spheres, determine each other. In fact, before Poincaré, Descartes studied the relations between the functions which defined the algebraic curves.

There are important *Riemann's theorem* (see Proposition 4.28) and *Riemann–Roch theorem* (see Proposition 4.34) for *Riemann surfaces* over complex numbers or complex algebraic curves. Those theorems are generalized to higher dimensions and to algebraic cases, say, over a finite field.

In coding theory, we use the *Riemann–Roch theorem* over a finite field. Hartshorne stated ([11], p. 293) that *if a reader is willing to accept the statement of the Riemann–Roch theorem, he can read this chapter at a much earlier stage of his study of algebraic geometry. This may not be a bad idea, pedagogically, because in that way he will see some applications of the general theory, and in particular will gain some respect for the significance of the Riemann–Roch theorem. In contrast, the proof of the Riemann–Roch theorem is not very enlightening.* What Hartshorne talked about was the algebraic geometry theory over an algebraically closed ground field. Our fields are not even algebraically closed, and the usual proofs in algebraic geometry over algebraically closed fields cannot be used without deep reflections and modifications. What we planned to do for *Riemann's theorem* (see the following) and *Riemann–Roch theorem* (see next section) are descriptive proofs, we give plenty examples and refer the hard parts of the proofs to Chevalley [10].

4.6.1. *Riemann's Theorem*

Before we state the theorem, let us consider the following examples.

Example 18: Let us consider the simplest curve, a straight line in the projective plane $P^2_\mathbf{K}$, defined by $x_2 = 0$. Let us consider the pole set $\{x_0 = 0, x_1 = 1, x_2 = 0\} = \{\mathbf{P}\}$. It is not hard to see that the set of all functions with at most one pole at $\mathbf{P}$ is $\{(a_0 x_0 + a_1 x_1)/x_0\}$ which is a vector space of dimension 2 over $\mathbf{K}$. In general, let $\mathbf{D} = n\mathbf{P}$. It is not hard to see that the set of all functions with at most $\mathbf{D} = n\mathbf{P}$ as poles are $\{(a_0 x_0^n + a_1 x_0^{n-1} x_1 + \cdots + a_n x_1^n)/x_0^n\}$ which is **a vector space of dimension** $n+1 = \ell(\mathbf{D})$ over $\mathbf{K}$. Let $t = x_1/x_0$. Then, $a_0 x_0^n + a_1 x_0^{n-1} x_1 + \cdots + a_n x_1^n)/x_0^n$ can be re-written as $a_0 + a_1 t + \cdots + a_n t^n$, a polynomial in t of degree $\leq n$. Then, we verify *Riemann's inequality* (cf. Proposition 4.28) $\ell(\mathbf{D}) \geq d(\mathbf{D}) + 1 - g$, with $g = 0$. ∎

Example 19: Assume that $ch(\mathbf{K}) \neq 2$ and $\mathbf{K}$ contains a square root of any element in $\mathbf{K}$. Let us consider another curve, an *elliptic curve* C in the projective plane, defined by $x_2^2 x_0 = (x_1 - ax_0)(x_1 - x_0)(x_1 - 0)$, where $a \neq 1, 0$. This projective curve is smooth (cf. Example 9). For the sake of our discussion, let us fix the affine piece $(A_K^2) = \{x_0 = 1\}$ as the points at finite distance. Let us consider the pole set $\{x_0 = 0 = x_1, x_2 = 1\} = \{\mathbf{P}\}$. Let us consider the set of all functions $L(n\mathbf{P})$ with at most n poles at $\mathbf{P}$, where n is a non-negative integer and no pole elsewhere. We **claim** that $L(n\mathbf{P})$ is a **vector space of dimension** n over $\mathbf{K}$.

We shall dehomogenize the equation by setting $x = \frac{x_1}{x_0}, y = \frac{x_2}{x_0}$. Then, the defining equation can be rewritten as

$$g(x,y) = y^2 - (x-a)(x-1)(x-0).$$

In general, the functions $f(x,y)$ in $F(C) = \mathbf{K}(x)[y]/(g(x,y))$ are of the following form:

$$f(x,y) = \frac{g_0(x) + g_1(x)y}{h(x)}.$$

We may assume that $(h(x), g_0(x), g_1(x)) = (1)$; otherwise, we may reduce the form. We make a further assumption that $h(x)$ splits completely (otherwise, go to a finite extension of $\mathbf{K}$, if the reader feels comfortable by assuming that $\mathbf{K}$ is algebraically closed, then assume it). Suppose that $f(x,y)$ has no pole at finite distance (means in A_K^2). We **claim** that $h(x)$ is a non-zero constant.

If not, we show that f has a pole at finite distance, and thus $f \notin L(n\mathbf{P}_\infty)$. Consider any non-constant factor $x - \beta$ of $h(x)$. We have the following two cases: (1) $\beta \neq 0, 1, a,$; (2) $\beta = 0, 1, a$. In case (1), the intersection of $x - \beta$ with curve C will be distinct points $\mathbf{P}_1$ and $\mathbf{P}_2$ (corresponding to two distinct non-zero values of y on C with $x = \beta$). Either $g_0(\beta) \neq 0$ or $g_1(\beta) \neq 0$ but not both are zero since $(h(\beta), g_0(\beta), g_1(\beta)) = 1$. In any situation, the numerator of $f(x,y)$ can not be 0 at both non-zero values of y. Therefore, f must have a pole at finite distance, and it is not allowed. In case (2), the intersection of $x - \beta$ with the curve C is at the point $\mathbf{P} = (\beta, 0)$ twice. If $g_0(\beta) \neq 0$ since $y = 0$, then the numerator does not pass through the point $\mathbf{P} = (\beta, 0)$, so the function will have two poles at $\mathbf{P} = (\beta, 0)$. It is impossible. Therefore, $g_0(\beta) = 0$. We shall study the completion of the local ring $O_\mathbf{P}$ at the point $\mathbf{P}$. Let us discuss the situation that $\beta = 0$ (other situations are similar). The defining equation is

$$y^2 = (x-a)(x-1)(x-0).$$

It is easy to see y is a uniformization parameter, and $\hat{O}_\mathbf{P} = \mathbf{K}[[y]]$. We have

$$x = a^{-1}y^2 + \cdots,$$
$$g_0 = b_2 y^2 + \cdots, \tag{1}$$
$$g_1 = c_0 + c_1 y + \cdots,$$

where $c_0 \neq 0$. It is easy to conclude that the numerator of f has $\mathbf{P}$ as zero once and the denominator of f has $\mathbf{P}$ as zero twice. Therefore, f has a pole at finite distance. Our claim is thus proved.

At $\mathbf{P}_\infty$, we set $u = \frac{x_0}{x_2} = y^{-1}, v = \frac{x_1}{x_2} = xy^{-1}$. The defining equation becomes

$$u = (v - au)(v - u)(v - 0).$$

It is easy to see that $\mathrm{ord}_{\mathbf{P}_\infty}(v) = 1, \mathrm{ord}_{\mathbf{P}_\infty}(u) = 3$. Hence. $\mathrm{ord}_{\mathbf{P}_\infty}(y) = -3$, and $\mathrm{ord}_{\mathbf{P}_\infty}(x) = -2$. Therefore, a polynomial of the form $g_0(x) + g_1(x)y$ will have order $-2i$ or $-2j - 3$ at $\mathbf{P}_\infty$. It is easy to see that there is no function with a simple pole at $\mathbf{P}_\infty$ and all functions with at most n poles at $\mathbf{P}_\infty$ form **a vector space of dimension** n, i.e., $\ell(n\mathbf{P}) = n$. In fact, the curve is of genus 1. We have the inequality of Riemann as

$$n \geq n + 1 - 1 = n + g - 1. \qquad \blacksquare$$

The shocking discovery is that the vector space $L(n\mathbf{P})$ is finite dimensional in general. We can tell the difference between two curves discussed in Examples 18 and 19 by studying $L(n\mathbf{P})$. This is one of Riemann's great discoveries.

There are many relations between $\ell(\mathbf{D})$ and $d(\mathbf{D})$. For instance, if $\mathbf{D} \succ 0$, it can be shown that $\ell(\mathbf{D}) \leq d(\mathbf{D}) + 1$. Let us prove it for a simple case that $\mathbf{D} = n\mathbf{P}$ with $\mathbf{P}$ a rational point, i.e., $\bar{O}_\mathbf{P} = \mathbf{K}[[t]]$. Let us define a map $\pi : L(\mathbf{D}) \mapsto \mathbf{K}^n$ as $\pi(f) = (a_1, \ldots, a_n)$, where $f \in L(\mathbf{D})$ and $f = a_n t^{-n} + \cdots + a_1 t^{-1} + \cdots$. Then clearly, π is a linear map and $\ker(\pi)$ $=\mathbf{K}$. Therefore, $\ell(\mathbf{D}) \leq n + 1 \leq d(\mathbf{D}) + 1$.

Similarly, it is easy to see that if $\mathbf{E} \succ \mathbf{D}$,i.e., $\mathbf{E} - \mathbf{D} = \mathbf{H} \succ 0$; then, $\dim(L(\mathbf{E})/L(\mathbf{D})) = \ell(\mathbf{E}) - \ell(\mathbf{D}) \geq d(\mathbf{E}) - d(\mathbf{D})$, i.e.,

$$\ell(\mathbf{E}) - d(\mathbf{E}) \geq \ell(\mathbf{D}) - d(\mathbf{D}).$$

Especially, if $\mathbf{E} = \mathbf{D} + \mathbf{P}$ for some rational point $\mathbf{P}$, then we have

$$\ell(\mathbf{E}) \leq \ell(\mathbf{D}) + 1.$$

Proposition 4.28. *We always have* $\ell(\mathbf{E}) - d(\mathbf{E}) \leq \ell(\mathbf{D}) - d(\mathbf{D})$ *for any two divisors* $\mathbf{E}$ *and* $\mathbf{D}$ *with* $\mathbf{E} \succ \mathbf{D}$.

Proof. Chevalley p. 21 [10]. □

We have the following Riemann's theorem. Note our notations are different from [10].

Proposition 4.29 (Riemann's theorem). *Let* X *be a given projective curve and* $\mathbf{D}$ *any divisor. Then, there is a minimal non-negative integer* g, *which will be called the* **genus** *of* X *such that*

$$\ell(\mathbf{D}) \geq d(\mathbf{D}) + 1 - g.$$

Moreover, if $d(\mathbf{D}) \geq 2g - 1$, *then*

$$\ell(\mathbf{D}) = d(\mathbf{D}) + 1 - g.$$

Proof. Chevalley p. 22 [10]. □

Proposition 4.30. *If* $d(\mathbf{D}) < 0$, *then* $\ell(\mathbf{D}) = 0$.

Proof. Let $f \in L(\mathbf{D})$. Then, we have $(f) + \mathbf{D} \succ 0$ and $d(\mathbf{D}) = d((f)) + d(\mathbf{D}) \geq 0$. A contradiction. □

4.6.2. *Plücker's Formula*

Note that the genus does not change under a separable field extension of the ground field $\mathbf{K}$ (cf. [10] Theorem 5, p. 99). We may extend the ground field $\mathbf{F}$ to its algebraic closure Ω without changing the genus. Hence, we shall mention the following classical *Plücker's formula* without the restriction that the ground field is algebraically closed.

Proposition 4.31 (Plücker's formula). *Let* C *be a smooth plane curve of degree* n. *Then, the genus* g *of* C *is given by the following formula:*

$$g = \frac{(n-1)(n-2)}{2}.$$ ∎

Remark: If the curve C is with only ordinary singularities $\mathbf{P}_i$ with multiplicities $\{m_i\}$ (i.e., a singularity with m_i distinct tangent lines), then the genus is

$$g = \frac{(n-1)(n-2)}{2} - \sum_i \frac{m_i(m_i - 1)}{2}.$$

It is called the extended *Plücker's formula*. ∎

Example 20: Let us consider the *elliptic curve* of **Example 19** with $n = 3$. It is easy to see that it is smooth and regular. Then, by Plücker's formula, its genus g is 1. It follows from Riemann's theorem that $\ell(\mathbf{D}) \geq d(\mathbf{D}) = n$ for $\mathbf{D} = n\mathbf{P}_\infty$, which has been verified by direct computation. ∎

Example 21: Let us consider a Fermat's curve $x_1^3 + x_2^3 + x_0^3$ over F_{2^m}. Then, by previous discussions, we know it is smooth and regular. By Plücker's formula, its genus g is 1. It follows from Riemann's theorem that $\ell(\mathbf{D}) \geq d(\mathbf{D})$ always. Let us verify Riemann's theorem for some special divisors.

Let $\mathbf{P}_\infty = (0, 1, 1)$. Let us consider $\mathbf{D} = n\mathbf{P}_\infty$ for some non-negative integer n. Note that $d(\mathbf{D}) = n$. Let us make a projective transformation π:

$$(1) : \pi(x_1) = y_1 + y_2,$$

$$(2) : \pi(x_2) = y_2,$$

$$(j) : \pi(x_0) = y_0.$$

Then, the defining equation becomes $y_1^3 + y_2 y_1^2 + y_2^2 y_1 + y_0^3$. Let us consider the affine part defined by setting $y_1 = 1$ with $x = y_0, y = y_2$. The equation becomes

$$y^2 + y = x^3 + 1.$$

The function field $F(C)$ is of degree 2 over $\mathbf{F}_{2^m}(x)$. In general, the functions in $F(C) = \mathbf{K}(x)[y]$ are of the following form:

$$f(x, y) = \frac{g_0(x) + g_1(x)y}{h(x)}.$$

We may assume that $(h(x), g_0(x), g_1(x)) = (1)$. We make a further assumption that $h(x)$ splits completely (otherwise, go to a finite extension of $\mathbf{K}$ or assume the field is algebraically closed). Suppose that $f(x, y)$ has no pole at finite distance, we claim that $h(x)$ is a non-zero constant.

If not, we show that f has a pole at finite distance and thus $f \notin L(n\mathbf{P}_\infty)$. Consider any non-constant factor $x - \beta$ of $h(x)$. The intersection of $x - \beta$ with curve C will be distinct points $\mathbf{P}_1 = (\beta, y_1)$ and $\mathbf{P}_2 = (\beta, y_2)$ (corresponding to two distinct non-zero values, y_1, y_2, of y on C with $x = \beta$, which can always be achieved if we go to an algebraic extension of $\mathbf{K}$). If the numerator of $f(x, y)$ is zero at both points, then we have

$$g_0(\beta) + y_1 g_1(\beta) = 0,$$

$$g_0(\beta) + y_2 g_1(\beta) = 0.$$

If we treat y_1, y_2 as numbers and $g_0(\beta), g_1(\beta)$ as variables, then we conclude

$$g_0(\beta) = 0,$$
$$g_1(\beta) = 0,$$

which contradicts our assumption that there is no common factor for $h(x), g_0(x), g_1(x)$! Our claim is thus proved.

At $\mathbf{P}_\infty = (0, 0, 1)$, we set $u = \frac{y_0}{y_2} = xy^{-1}, v = \frac{y_1}{y_2} = y^{-1}$. The defining equation becomes

$$v^2 + v = u^3 + v^3.$$

It is easy to see that $\mathrm{ord}_{\mathbf{P}_\infty}(v) = 3, \mathrm{ord}_{\mathbf{P}_\infty}(u) = 1$. Hence, $\mathrm{ord}_{\mathbf{P}_\infty}(y) = -3$, and $\mathrm{ord}_{\mathbf{P}_\infty}(x) = -2$. Therefore, a polynomial of the form $g_0(x) + g_1(x)y$ will have order $-2i$ or $-2j - 3$ at $\mathbf{P}_\infty$. It is easy to see all functions with at most n poles at $\mathbf{P}_\infty$ form a vector space of dimension n. ∎

Example 22: Let us consider the curve $x^2 - y^2 + x^3 + y^4$ defined over the complex numbers $\mathbb{C}$. By the Jacobian criteria, the origin is the only singularity with ordinary double multiplicity. Hence, by an **extended** Plücker formula, which works for singular curves, its genus g is $\frac{(4-1)(4-2)}{2} - 1 = 2$ (cf. [15]).

From the above, we easily conclude that a smooth projective plane curve of degree n is of genus $(n - 1)(n - 2)/2$. Therefore, let $n = 1, 2, 3, 4, \ldots$, the genera are $0, 0, 1, 3, \ldots$. If there is a plane curve of genus 2, then there must be some singular points in any its planar model. ∎

If we consider all algebraic projective curves, the coarsest classification is by their *genera*, i.e., two curves are in the same class if their genera are the same. This is the *discrete* parameter the classification of algebraic smooth projective curves according to their genera. If the ground field is the complex number field $\mathbb{C}$, then it simply states the underlying topological spaces are the same for all algebraic projective curves with same given genus. However, *analytically*, the structure can still be different. Thus, we have a further fine parameter for all algebraic projective curves over the complex field $\mathbb{C}$ with the same genus, namely, the variety of moduli of curves of genus g, which is a point if $g = 0$, has dimension 1 if $g = 1$ and dimension $3g - 3$ if $g \geq 2$.

4.6.3. *Rational Curve*

A curve C is said to be a *rational curve* if the rational function field of C is isomorphic to $K(x)$. We have the following proposition.

Proposition 4.32. *Let* $\mathbf{F}$ *be an algebraic function field of one variable. Then,* $\mathbf{F} \equiv \mathbf{K}(x) \Leftrightarrow C$ *has a rational point and* $g = 0$.

Proof. Chevalley p. 23 [10]. $\square$

Example 23: Let C be a plane curve over $\mathbf{K} = \mathbf{F}_q(u, v)$, where u, v are symbols, defined by

$$ux^m + vy^m = 1,$$

with $m \geq 2, p \nmid m$. It follows from Plücker's formula that the genus of C is $(m - 1)(m - 2)/2$. We claim that there is no rational point. If there is one, let its coordinate be $(g(u, v)/f(u, v), h(u, v)/f(u, v))$, where $f(u, v), g(u, v), h(u, v)$ are polynomials in u, v; we may further assume that there is no common factor among all three polynomials, then the above equation can be rewritten as

$$ug(u, v)^m + vh(u, v)^m = f(u, v)^m.$$

Let $\deg(f(u, v)) = a, \deg(g(u, v)) = b, \deg(h(u, v)) = c$. Then, the degrees of those three terms are $1 + mb, 1 + mc, ma$, respectively. Moreover, the highest two of them must be equal. We conclude that the highest forms of $ug^m(u, v)$ and $vh^m(u, v)$ must have factors in v of degrees ma' and $mb' + 1$, then the terms cannot cancel out. This is impossible.

Since there is no rational point, for $m = 2$, the curve C has genus $g = 0$ but is not a projective line. $\blacksquare$

Exercises

(1) Show that the curve defined over $\mathbf{F}_{p^2}$ by the equation

$$ax^2 + by^2 = 1$$

is a birationally equivalent (cf. Definition 4.26) to a projective line for any $0 \neq a, b \in F_{p^2}$, where p is an odd prime number.

(2) Finish the arguments of **Example 22**.

(3) Finish the arguments of **Example 23**.

(4) Prove Riemann's theorem for P_K^1 directly.

(5) Given $ch(\mathbf{K}) > 2$, and the curve C defined by $x^2 - y^2 + x^3 + y^4 = 0$, let $\mathbf{P}$ be the point at infinity. Find $L(\mathbf{P}), L(2\mathbf{P}), L(3\mathbf{P})$.

4.7. Riemann–Roch Theorem I

Note that Riemann's theorem is an inequality. There is a number missing for it to be an equality. Roch, a student of chemistry who had a Ph.D. in electromagnetism and worked under Riemann, found a natural explanation of the missing number. Riemann passed away on July 20, 1866 at the age of 40, and Roch, a few months later, died at the age of 26. Their work has been generalized to functions of one variable over a finite field which is useful for coding theory. First, we have to introduce the abstract concepts of *differentials*. We shall follow Chevalley [10] in our treatment of *Riemann–Roch* theorem. It gives us a fast way of proving *Riemann–Roch* theorem. We quote [10] to show that the abstractly defined term *differential* in the next section is really the ordinary *differential* in the classical analysis. Classically, we have the following interesting remark.

Remark: Let us consider the complex case, where $\mathbf{K} = \mathbb{C}$ the field of complex numbers. Let $\omega = f(x)dx$ be any classically named *differential* and $g(x)$ be any meromorphic function locally defined at a point $\mathbf{P}_j$. Then, we may treat ω as a linear functional on the meromorphic function $g(x)$, locally defined at $\mathbf{P}_j$ as the *(Cauchy) residue of $g(x)f(x)dx$ at a point $\mathbf{P}_j$*:

$$\omega(g(x)) = \frac{1}{2\pi i} \int_\Gamma g(x)f(x)dx$$
$$= \textit{the} \text{ residue of } g(x)f(x)dx \text{ at a point } \mathbf{P}_j,$$

where Γ is a simple loop around $\mathbf{P}_j$ such that $g(x)f(x)dx$ has no pole on the loop nor other poles inside the loop and $i = \sqrt{-1}$ as usual in complex analysis. Note that $\omega(g(x)) \in \mathbb{C}$, and ω can be considered as a linear function in $g(x)$. Furthermore, if we take the loop small enough so that there is no pole inside, then the integral is 0. Therefore, we take the same loop with the reverse direction, then the integration is still 0, which means $\sum \mathrm{res}_{\mathbf{P}_j}(g(x)f(x)dx) = -0 = 0$ for global rational function $g(x)$. That is an important classical Cauchy theorem: $\sum \mathrm{res}_{\mathbf{P}_j}(g(x)f(x)dx) = \sum \mathrm{res}_{\mathbf{P}_j}(h(x)dx) = 0$.

   Therefore, we may treat *differentials* as functionals on *repartitions* (see Definition 4.32). Furthermore, the *residue* which was defined by integration can be defined pure algebraically as with x a *local uniformizing parameter* and given the (meromorphic) expression of $h(x) = g(x)f(x)$ as

$$h(x) = \sum_{j=-m}^{\infty} a_j x^j.$$

Then, we may simply define residue as a_{-1} and disregard the integration. Certainly, we have to prove the residue thus defined is algebraically sound. ∎

4.7.1. *Repartitions and Differentials*

In the above remark, we use an integration theory. In our present situation, we are without the usage of integrations. Weil found the correct algebraic way of treating differentials. We define *repartitions* and *differentials* for the general cases as follows (Chevalley [10]).

Definition 4.33. A *repartition* ξ is an assignment to every place **P** a function $\xi_{\mathbf{P}}$ such that there are only finitely many places **P** with $\operatorname{ord}_{\mathbf{P}}(\xi_{\mathbf{P}}) < 0$. The collection of all *repartitions* is denoted by Ξ. Note that a rational function f has only finitely many poles $\mathbf{P}_j$. It is clear that we may assign the rational function f to all places $\mathbf{P}_j$. It is called the *constant repartitions*. The collection of constant repartitions is isomorphic to **F** and will be denoted by the same symbol **F**. Given any divisor **D**, we define $\Xi(\mathbf{D})$ as the collection of all repartition ξ such that

$$\operatorname{ord}_{\mathbf{P}}(\xi_{\mathbf{P}}) + \operatorname{ord}_{\mathbf{P}}(\mathbf{D}) \geq 0$$

for all **P**. ∎

We give the following general definition, (cf. [10] p. 30).

Definition 4.34. A *differential* ω is a (**K**-)linear functional on Ξ such that there exists a divisor **D** with ω vanishing identically on $\Xi(-\mathbf{D}) + \mathbf{F}$, where $\Xi(-\mathbf{D})$ is defined in the previous definition. If **D** is any divisor which has the stated property, then we say the differential ω is a multiple of **D**, and we write $\omega \equiv 0(\mod (\mathbf{D}))$ and $\omega \in \Omega(\mathbf{D})$.

Let us define the divisor $\delta(\omega)$ of a differential ω as the maximal **D** such that $\omega \in \Omega(\mathbf{D})$. Clearly, such a divisor exists and is unique (cf. [10] p. 32. Note that Chevalley used a^n with a an ideal and considered a^n maximal). We may reformulate $\Omega(\mathbf{D}) = \{\omega : \delta(\omega) \succ \mathbf{D}\}$. We define the degree $\ell(\omega) = \ell(\delta(\omega))$ and $d(\omega)$ as $d(\delta(\omega))$, the divisor of a differential is called a *canonical divisor*.

Furthermore, we name $i(\mathbf{D})$ the *index* of **D** to denote the dimension of the vector space $\Omega(\mathbf{D})$. ∎

4.7.2. Riemann–Roch Theorem

The above definitions are abstract and general. They provide a short proof for a form of the Riemann–Roch theorem we need.

Let R be the field of algebraic functions of one variable of genus g, and let $\mathbf{D}$ be a divisor of R. We have the following Riemann–Roch theorem.

Proposition 4.35 (Riemann–Roch theorem). *Let us assume the preceding paragraph. We have*

$$\ell(\mathbf{D}) - i(\mathbf{D}) = d(\mathbf{D}) + 1 - g.$$

Proof. Chevalley, p. 30 [10]. $\qquad\qquad\qquad\qquad\qquad\qquad\qquad\qquad$ □

4.7.3. Canonical Class

Let $\mathbf{F}$ be a function field of one variable over a perfect field $\mathbf{K}$. Let $\omega \neq 0$ be a differential of $\mathbf{F}$ over $\mathbf{K}$. Then, we have the following proposition.

Proposition 4.36. *The vector space of all differentials can be written as $\mathbf{F}\omega$.*

Proof. Let $\omega' \in \mathbf{F}\omega$. Then, ω' is a differential (see the paragraph before Theorem 5 on p. 31 of [10]). Furthermore, we refer to Theorem 5 on p. 31 of [10] to see that any differential ω' can be written as $x\omega$ with $x \in \mathbf{F}$. □

We have the following proposition about the important numbers $d(\omega)$ and $\ell(\omega)$.

Proposition 4.37. *Let $\mathbf{W}$ be a canonical divisor. Then, we have the following:* (1) *All canonical divisors $\mathbf{W}$ are linearly equivalent.* (2) $d(\mathbf{W}) = 2g - 2$. (3) $\ell(\mathbf{W}) = g$, *where g is the genus of the curve C.*

Proof. (1) It is easily provable. (2) Chevalley, p. 32 [10]. (3) Chevalley, p. 30 [10]. $\qquad\qquad\qquad\qquad\qquad\qquad\qquad\qquad\qquad\qquad\qquad\qquad$ □

It follows from the preceding proposition that if we know one non-zero differential ω, then we know $\mathbf{F}\omega$ and all differentials. What we plan to do in the next section is to show that we can define dx as a non-zero differential.

4.7.4. Residue

In the example at the beginning of this section, we have the important concept of *residue* of Cauchy over complex numbers $\mathbb{C}$. We want to generalize the concept to the cases of finite fields. Algebraically, we define the residue field of a valuation O as $O/\mathbf{P}$, where $\mathbf{P}$ is the only maximal ideal of O. In our present situation, we have $\mathbf{K}$ a perfect field, hence $O/\mathbf{P}$ is separable over $\mathbf{K}$. Any differential ω when restricted to a place $\mathbf{P}$ will induce a linear map on $O/\mathbf{P}$ as a finite-dimensional vector space over $\mathbf{K}$. It can be proved that any linear map of $O/\mathbf{P}$ can be represented as $a :\mapsto trace(a)\sigma$, where $\sigma \in O/\mathbf{P}$ is uniquely determined by the map (see [10], Lemma 1, p. 48). We define the residue of ω to the unique element σ. Although the definition of *residue* looks strange, it is really the generalization of Cauchy's definition by integration! However, we use the *residues* in the coding theory only for rational points of the curve where the computations of residues of *differentials* are simple (see Proposition 4.42).

Similar to the classical analysis, we have the following interesting proposition.

Proposition 4.38. *Let ω be a differential with poles $\{\mathbf{P}_j\}$, then we have*

$$\sum \mathrm{res}_{\mathbf{P}_j}(\omega) = 0.$$

Proof. Chevalley, p. 49 [10]. □

Proposition 4.39 (The existence theorem for differentials). *Let $\mathbf{P}_1, \ldots, \mathbf{P}_m$ be finitely many distinct rational points. Let $\mathbf{D} = \sum \mathbf{P}_i$. Given elements $r_i, \ldots, r_m$ in $\mathbf{K}$ such that*

$$\sum_i r_i = 0.$$

Then there exists a differential ω of $\mathbf{F}$ with the divisor $(\omega) \in \Omega(\mathbf{D})$ such that $\mathrm{res}_{P_i} \omega = r_i$.

Proof. Chevalley, p. 50 [10]. □

Remark: If there is another rational point $\mathbf{P}'$ outside the set $\{\mathbf{P}_1, \ldots, \mathbf{P}_m\}$, then we may push the residue to $\mathbf{P}'$, i.e., define $r' = -\sum_i r_i$, then the condition $r' + \sum_i r_i = 0$ is automatically satisfied. Let $\mathbf{D} = \sum \mathbf{P}_i$. Note that $\Omega(\mathbf{D} + \mathbf{P}') \subset \Omega(\mathbf{D})$. ∎

Exercises

(1) Prove the *existence theorem for differential* for P_K^1 directly.

(2) Prove the *approximation theorem for functions* for P_K^1 directly.

4.8. Riemann–Roch Theorem II

4.8.1. *The differential dx for* $\mathbf{K}(x)$ *and* $\mathbf{F}$

From the Proposition 4.35 of the last section, we know that it suffices to find just one non-zero differential ω, then we know all other differentials. For this purpose, let us select $x \in \mathbf{F}$ such that $\mathbf{F}$ is finite separable over $\mathbf{K}(x)$ (it can be done by Noether's theorem). We shall first select a special differential dx for $\mathbf{K}(x)$ and then extend dx to $\mathbf{F}$. We have the following proposition.

Proposition 4.40. *Let us consider the field* $\mathbf{K}(x)$ *as selected in the preceding paragraph. Let* $\mathbf{P}_\infty$ *be the place at* ∞ *and* $\mathbf{D}_j = -j\mathbf{P}_\infty$. *We have* $\ell(\mathbf{D}_j) = 0$ *for* $j = 1,2$ *and* $i(\mathbf{D}_1) = 0, i(\mathbf{D}_2) = 1$.

Proof. Since the degrees of $\mathbf{D}_j$ are negative, it follows from Proposition 4.29 that we have $\ell(\mathbf{D}_j) = 0$. The second part of the proposition follows from Riemann–Roch theorem since $g = 0$. $\qquad\square$

Let us use the material on p. 102 in [10] for the following proposition.

Proposition 4.41. *There is a unique differential* $\omega \in \Omega(\mathbf{D_2})$ *which has value* -1 *on* $\xi = \frac{1}{x}$ *at* $\mathbf{P}_\infty$. $\qquad\blacksquare$

We shall denote the unique differential by dx, or more precisely by $dx_{K(x)}$. Consider $\mathbf{F} \supset \mathbf{K}(x)$ and a separable extension of $\mathbf{K}(x)$. We shall use the usual method to extend $dx_{K(x)}$ to $dx_{\mathbf{F}}$. We have

$$dx_{\mathbf{F}} = cotrace_{K(x)/\mathbf{F}}(dx_{K(x)}),$$

where *cotrace* is defined on p. 105 in [10] as

$$cotrace_{K(x)/\mathbf{F}}(\omega)(\xi) = \omega(trace_{\mathbf{F}/K(x)}(\xi)),$$

while *trace* is the usual one. We may abuse the notation and use dx for the field $dx_{\mathbf{F}}$. The only thing we have to know is the value of $(dx)y = ydx$ at a place $\mathbf{P}$; we define it to be the *residue* of ydx at $\mathbf{P}$ (see the following).

4.8.2. *Complete Local Ring*

We shall quote the following theorem of Cohen (cf. [16] p. 304).

Proposition 4.42. *An equicharacteristic complete local ring* $\overline{O}$ *admits a field of representatives.* $\qquad\blacksquare$

The above proposition means that in our situation, a local ring O (we only consider valuation rings) contains a perfect field $\mathbf{K}$, hence the residue map $O \mapsto O/\mathbf{P} = \overline{O}$ induces an isomorphic on $\mathbf{K}$. So, the characteristic of $\mathbf{K}$ equal to the characteristic of $\overline{O}$. This is the equicharacteristic case, and $\overline{O}$ is $\mathbf{K}'[[t]]$, where $\mathbf{K}'$ is the field of representative. We have the following proposition which will be useful for *algebraic coding theory*.

Proposition 4.43. *Let C be a smooth projective curve, and let $\mathbf{P}$ be a rational point, i.e., $(O/\mathbf{P}=\mathbf{K})$, and uniformization parameter $t \in R$ and xdy be a (classical) differential. Let ydx be expressed as*

$$ydx = \left(\sum_{r}^{\infty} c_j t^j \right) dt \in \mathbf{K}((t))dt \qquad c_r \neq 0.$$

Then, we have the following: (1) $\mu_p(ydx) = r$; (2) $\mathrm{res}_\mathbf{P}(ydx) = c_{-1}$.

Proof. Statement (1) follows from definition, and for (2), see Chevalley, p. 110 corollary of Theorem 6 [10]. □

Remark: In our applications to coding theory, we restrict the above proposition to the simple cases that the differentials only have simple poles, i.e., $r \geq -1$.

Note that in the classical complex case, every place is rational; this result shows that this residue matches Cauchy residue.

Example 24: Let us consider the curve $C = P_\mathbb{C}^1$, where $\mathbb{C}$ is the field of the complex numbers.

(1) We shall make computations according to Riemann–Roch theorem. Let $\mathbf{D} = m\mathbf{P}_\infty$, where m is an integer ≥ -1. Since the genus $g = 0$, we have $d(m\mathbf{P}) \geq 2g - 1 = -1$; it follows from Proposition 4.28 (Riemann's theorem) that $\ell(m\mathbf{P}) = d(m\mathbf{P}) + 1 - g$. Hence, it follows from Proposition 4.34 (Riemann–Roch theorem) that $i(m\mathbf{P}) = 0$, i.e., there is no differential ω with divisor $\delta(\omega) \succ m\mathbf{P}_\infty$. Let us consider $\mathbf{D} = -m\mathbf{P}_\infty$, where $m \geq 2$ is an integer. Let the regular function ring of $P_\mathbb{C}^1 \backslash \{0\}$ be $\mathbb{C}[x]$. Then, it follows from Proposition 4.29 that $\ell(-m\mathbf{P}) = 0$. Further, it follows from Proposition 4.34 (Riemann–Roch theorem) that

$$-i(-m\mathbf{P}) = d(-m\mathbf{P}) + 1 - g,$$

and hence, we have

$$i(-m\mathbf{P}) = m - 1.$$

Let $\frac{f(x)}{g(x)}dx \in \Omega(-m\mathbf{P})$. Since the differential has no pole in $A_{\mathbb{C}}^1$, $g(x)$ must be a non-zero constant; we may assume it is 1, i.e., the differential is $f(x)dx$. It is easy to check that the vector space $\Omega(-m\mathbf{P})$ is generated by $\{dx, xdx, \ldots, x^{m-2}dx\}$.

(2) We wish to show that the *differentials* in the classical sense are equivalent to linear functions on Ξ. First, we generalize our previous concepts as follows.

Let $f(x)dx$ be a differential in the classical sense. Let $v_p =$ the order of $f(x)dx$ at point $\mathbf{P}$, i.e., $f(x)dx = (\sum_{i=v_p}^{\infty} a_j t^j)dt$, where $a_{v_p} \neq 0$ and t is a uniformization parameter. Then, $(f(x)dx) = \sum v_p \mathbf{P}$. Let $\mathbf{D} = \sum_j n_j \mathbf{P}_j$ be a divisor. Then, we have $f(x)dx \in \Omega(\mathbf{D}) \iff (f(x)dx) \succ \mathbf{D}$. We give a concrete argument for (2) directly. We separate the discussions into the following two steps: (A) Every differential in the classical sense induces a differential in the Weil's sense, and this induction is a one to one correspondence. (B) The induced map is a bijection between $\Omega(\mathbf{D})$ and the linear functionals vanishing on $\Xi(\mathbf{D}) + \mathbf{F}$.

Proof of (A): We show that every *differential* in the classical sense can be viewed as a ($\mathbf{K}$-)linear function on Ξ which vanishes on $\Xi(\mathbf{D}) + \mathbf{F}$ for some suitable divisor $\mathbf{D}$. Let $\omega = f(x)dx$ be a differential in the classical sense. Then, for any *repartition* ξ and any point $\mathbf{P} \in P_C^1$ (note that all points are rational), let ξ_p be the function element specified by ξ at the point $\mathbf{P}$, we may define

$$\omega_{\mathbf{P}}(\xi) = the \ residue \ of \ \xi_p \omega \ at \ \mathbf{P}.$$

Since for the above residue to be non-zero at $\mathbf{P}$, either ω must have a pole or $\xi_{\mathbf{P}}$ must have a pole, and both sets are finite, therefore there are only finitely many $\mathbf{P}$ for the above residue to be non-zeroes, we may define

$$\omega(\xi) = \sum_{\mathbf{P}} \omega_{\mathbf{P}}(\xi).$$

From the above, we conclude that ω defines a linear function on Ξ. It is clear that for any constant repartition $\xi = g(x) \in \mathbf{F}$, $\omega(\xi) = \sum \text{res}(g(x)f(x)dx) = 0$.

For any differential $f(x)dx$ in the classical sense, let $(f(x)dx) = \sum_i n_i \mathbf{P}_i - \sum_j m_j \mathbf{P}_j$, where n_i, m_j are non-negative; let $\mathbf{D} = \sum_i n_i \mathbf{P}_i - \sum_j m_j \mathbf{P}_j$. It is easy to see that $(f(x)dx) = \mathbf{D}$, $(f(x)dx) \succ \mathbf{D}$ and $(f(x)dx) \in \Omega(\mathbf{D})$. Furthermore, $\omega_{\mathbf{P}}(\xi) = 0$ for all $\mathbf{P}, \xi \in \Xi(\mathbf{D})$. Therefore, ω vanishes on $\Xi(\mathbf{D}) + \mathbf{F}$. We define a map from the differentials in the classical sense $\in \Omega(D)$ to elements in the new sense which vanish on $\Xi(D) + \mathbf{F}$.

Moreover, we claim that any non-zero differential $\omega = f(x)dx$ in the classical sense cannot induce a zero-linear function. Let a be any point that is neither a zero nor a pole of $f(x)$. Consider a *repartition* ξ defined by $\xi = (x - a)^{-1}$ at $\mathbf{P} = (x - a)$ and 0 elsewhere. Then, $\omega(\xi) = f(a) \neq 0$.

Proof of (B): We wish to show that the induced residue map is onto. We observe that the classical differential $\omega \in \Omega(\mathbf{D})$ will vanish on $\Xi(\mathbf{D}) + \mathbf{F}$ for any given divisor $\mathbf{D}$. Therefore, the induced map by the residue will send $\Omega(\mathbf{D})$ to a subspace of all linear functionals which vanish on $\Xi(\mathbf{D}) + \mathbf{F}$. We wish to show that they have the same dimensions. Then, clearly it will imply that the induced residue map is onto. We shall count the number of linear independent classical *differentials* in $\Omega(-\mathbf{D})$ for any divisor $\mathbf{D}$. We claim that the number is $\max\{0, d(\mathbf{D}) - 1\}$.

Let $-\mathbf{D} = \mathbf{D_A} + \mathbf{D_\infty}$, where $\mathbf{D_A}$ is the affine part of $-\mathbf{D}$ and $\mathbf{D_\infty}$ is the part of $-\mathbf{D}$ at ∞. Consider $\mathbf{D_A} = \sum n_j \mathbf{P}_j$ with $\mathbf{P}_i = (x - a_j)$ and $\mathbf{D_\infty} = m\mathbf{P_\infty}$. Let $h(x) = \prod_j (x - a_j)^{-n_j}$ and $f(x)dx \in \Omega(-\mathbf{D})$, i.e.,

$$(f(x)dx) \succ -\mathbf{D}.$$

Then, $(f(x)dx) = \sum_j n'_j \mathbf{P}_j + m'\mathbf{P_\infty} + \sum_k \ell_k \mathbf{P}_k$ with $n'_j \geq n_j$, $m' \geq m$ and $\ell_k \geq 0$. We see that $h(x)f(x)$ has no poles at finite distance, and thus, it is a polynomial $g(x)$ in x. Therefore $(g(x)dx)$ is $\sum(n'_j - n_j)\mathbf{P}_i + (m' + (\sum_j n_j)\mathbf{P_\infty}$. Let $t = x^{-1}$ be a uniformization parameter at ∞. We have $dx = -t^{-2}dt$. Then, it follows that at the point ∞,

$$v_\infty(h(x)f(x)dx) = m' + \left(\sum_j n_j\right) \geq m + \left(\sum_j n_j\right) = -d(\mathbf{D}),$$

$$v_\infty(g(x)) \geq 2 - d(\mathbf{D}).$$

It means that $g(x)$ is a polynomial of degree $\leq d(\mathbf{D}) - 2$. The set of such $g(x)$ forms a vector space of dimension $d(\mathbf{D}) - 1$ or 0.

We claim that the number of linear-independent linear functions on Ξ which vanish on $\Xi(-\mathbf{D}) + \mathbf{F}$ is $\max\{0, d(\mathbf{D}) - 1\}$.

Let us discuss a simple case, say, $\mathbf{D} = \mathbf{P} \neq \mathbf{P_\infty}$. Let ξ be any repartition. It is easy to see that there is a rational function $f(x)$ such that $\xi - f(x)$ has no pole. Let $\xi_\mathbf{P}(\mathbf{P}) = a$ which may be 0. We shall consider $\xi - f(x) - a$, then this repartition has a zero at $\mathbf{P}$; therefore, it is in $\Xi(-\mathbf{P})$ and $\xi \in \Xi(-\mathbf{P}) + \mathbf{F}$. We conclude that a linear function which vanishes on $\Xi(-\mathbf{P}) + \mathbf{F}$ must vanish on any repartition ξ, which means the differential which vanishes on $\Xi(-\mathbf{P}) + \mathbf{F}$ must be the zero-linear function. Our claim is proved in this case. For the general cases, the reader is referred to Chevalley, pp. 26–30 [10] or Exercise 4.

According to $(1), (2), (3)$, *differentials* in the classical sense are the linear functions which vanish on $\Xi(\mathbf{D}) + \mathbf{F}$. ∎

Example 25: Let $C = P^1_K$ with $ch(\mathbf{K}) > 2$. Then, $F(C) = \mathbf{K}(x)$. Let $\omega = (x^2 + x^3)d(x^2) + dx$ be a differential. It is easy to see that $\omega = (1 + 2x^3 + 2x^4)dx$. At any point $\mathbf{P}_a = ((x - a))$, the local ring at $\mathbf{P}_a$ is $R = \mathbf{K}[x]_{(x-a)}$ with maximal ideal $(x-a)R$, where $(x-a)$ is the uniformization parameter. Then, $\omega = (1 + 2(x - a + a)^3 + 2(x - a + a)^4)d(x - a) = f(x - a)d(x - a)$. We have $\mathrm{ord}_{\mathbf{P}_a}(\omega) = \mathrm{ord}_{\mathbf{P}_a}(f(x - a))$. We have one more point $\mathbf{P}_\infty$ at ∞. The local ring at $\mathbf{P}_\infty$ is $\mathbf{K}[\tau]_{(\tau)}$, where $\tau = \frac{1}{x}$. The differential $\omega = (1 + 2\tau^{-3} + 2\tau^{-4})d(\tau^{-1}) = -(\tau^{-2} + 2\tau^{-5} + 2\tau^{-6})d\tau) = -\tau^{-6}(2 + 2\tau + \tau^4)d\tau$. So, we have that $\mu_{\mathbf{P}_\infty}(\omega) = \mathrm{ord}_{\mathbf{P}_\infty}(-(\tau^{-2} + 2\tau^{-5} + 2\tau^{-6}) = -6$. ∎

Example 26: In the definition of *residue*, we have to use the *trace* function which can be illustrated by examples. If the point is not rational, then we have to consider the *trace* function. Let us consider the curve P^1_K and the following differential:

$$\eta = \frac{2x\,dx}{x^2 + 1}.$$

Let the ground field $\mathbf{K}$ be $\mathbb{C}$ the complex field. By partial fraction, we have with $i = \sqrt{-1}$,

$$\eta = \left(\frac{1}{x + i} + \frac{1}{x - i} \right) dx.$$

It is easy to see that the residues at $(x + i), (x - i)$ are $1, 1$, respectively. At $\mathbf{P}_\infty$ with uniformization parameter $t = x^{-1}$ and $x^2 + 1$ is an unit, η becomes

$$\eta = (-2t^{-1} + \cdots)dt.$$

It is easy to see that $\sum \mathrm{res}_{\mathbf{P}_j} \eta = 0$. However, let us consider the case that the ground field $\mathbf{K} = \mathbb{R}$ the real field. At ∞, it has a residue -2 as before. There is only one point $\mathbf{P} = (x^2 + 1)$ at finite distant with a pole. Let $t = x^2 + 1$ be a uniformization parameter at this point. Then, we have

$$\eta = t^{-1}dt.$$

The coefficient of t^{-1} term is 1. However, the point is not rational. Its residue will be defined as the *trace* of 1, note that $1 \cdot 1 = 1 \cdot 1 + 0 \cdot i$ and

$1 \cdot i = 0 \cdot 1 + 1 \cdot i$. It is easy to deduce that

$$\text{res}_\mathbf{P}(\eta) = Tr_{\mathbb{C}/\mathbb{R}}(I) = 2,$$

where $Tr_{\mathbb{C}/\mathbb{R}}$ is the *trace* operation from $\mathbb{C}$ to $\mathbb{R}$. We still have $\sum \text{res}_{\mathbf{P}_i} \eta = 0$.
Let us consider another differential ω defined as

$$\omega = \frac{2dx}{x^2 + 1}.$$

Let the ground field $\mathbf{K}$ be $\mathbb{C}$ the complex field. By partial fraction, we have
with $i = \sqrt{-1}$,

$$\omega = \left(\frac{i}{x + i} + \frac{-i}{x - i} \right) dx.$$

It is easy to see that the residues at $(x+i), (x-i)$ are $i, -i$, respectively.
At $\mathbf{P}_\infty$ with uniformization parameter $t = x^{-1}, dt = -x^{-2}dx + \cdots$. The
differential ω becomes

$$\omega = (-2 + \cdots)dt.$$

Hence, the residue is 0 at ∞. It is easy to see that $\sum \text{res}_{\mathbf{P}_j} \omega = 0$. However,
let us consider the case that the ground field $\mathbf{K} = \mathbb{R}$ the real field. At ∞, it
has a residue 0. There is only one point $\mathbf{P} = (x^2 + 1)$ at finite distant with
a pole. Let $t = x^2 + 1$ be a uniformization parameter at this point. Then
we have **Example 7** in Section 4.3 where we show that $x = i\sqrt{1 - t}$ in the
complete local ring $\mathbb{C}((t))$:

$$\omega = (x(t^{-1}))dt = (it^{-1} + \cdots)dt.$$

The coefficient of t^{-1} is i. Its residue will be defined as the *trace* of $i(= \sqrt{-1})$. Note that $i \cdot 1 = 0 \cdot 1 + 1 \cdot i$ and $i \cdot i = -1 \cdot 1 + 0 \cdot i$; it is easy to
deduce

$$\text{res}_\mathbf{P}(\omega) = Tr_{\mathbb{C}/R}(i) = 0 + 0 = 0,$$

where $Tr_{\mathbb{C}/R}$ is the *trace* operation from $\mathbb{C}$ to $\mathbb{R}$. We still have $\sum \text{res}_{\mathbf{P}_j} \eta = 0$.
∎

Example 27: Let us consider the algebraic curve P_K^1. Let $\eta = f(x)dx = \frac{r(x)}{s(x)}dx$ be any non-zero differential. For simplicity, we assume that $s(x)$ can
be factored into linear polynomials (otherwise, go to a finite field extension
of $\mathbf{K}$). By the theory of partial fractions, we have

$$\eta = (u(x) + \sum_{j,k,a_j} \frac{c_{jk}}{(x - a_j)^k})dx.$$

The sum of all residues at finite distance is clearly $\sum c_{j1}$. We shall show that the residue at $\mathbf{P}_\infty$ is its negative. Since the residue map is linear, it suffices to check every term in the above formula. Note that at $\mathbf{P}_\infty$, a uniformizing parameter is $x^{-1} = t$. Then, we have

$$u(x)dx = -t^{-2}u(t^{-1})dt,$$

$$\left(\frac{c_{jk}}{(x-a_j)^k}\right)dx = -t^{-2}\left(\frac{c_{jk}t^k}{(1-a_jt)^k}\right)dt, \quad \text{if } k \geq 2,$$

$$\left(\frac{c_{j1}}{(x-a_j)}\right)dx = -t^{-2}\left(\frac{c_{j1}t}{(1-a_jt)}\right)dt.$$

Clearly, the residue at $\mathbf{P}_\infty$ is $-\sum c_{j1}$ and combining with our previous result that the sum of all residues at finite distance is clearly $\sum c_{j1}$, we have $\sum \mathrm{res}_{p_j}(\eta) = 0$.   ∎

Example 28: Let us consider **Example 19** of the preceding section. As usual, we assume that the field $\mathbf{K}$ is not of characteristic 2. We have an equation which defines a projective curve C of genus 1 with $a \neq 0, 1$:

$$y^2z = (x - az)(x - z)(x - 0z).$$

Let us dehomogenize the above equation by set $z = 1$; the equation becomes $y^2 = (x-a)(x-1)(x-0)$. Let us consider the differentials dx, dy. By implicit differentiation, we have

$$dy = \left(\frac{3x^2 - 2(1+a)x + a}{2y}\right)dx.$$

Therefore, dx and dy are linearly equivalent. Let us consider the differential dx. We consider the curve C at finite distance as a *covering* of the affine line $\mathbf{A}_{\mathbf{K}}^1$. At finite distance of the affine line $\mathbf{A}_{\mathbf{K}}^1$, let us consider the following:(1) a rational point $(x-\beta)$ with $\beta = 0, 1, a$; (2) a rational point with $\beta \neq 0, 1, a$; (3) non-rational point $f(x)$, where $f(x)$ is a monic irreducible polynomial of degree > 1 as follows.

(1) Our defining equation provides $y = 0$ and y is a uniformization parameter. Then, we have $x = y^2 + \cdots$ and $dx = (2y + \cdots)dy$. Therefore, dx has a 0 of order 1 at these three points.

(2) The value of y must be $\pm\sqrt{(\beta-a)(\beta-1)(\beta)} = \pm c$, where c may or may not be in $\mathbf{K}$. We separate the discussions into following two subcases:

    (A) Suppose that $0 \neq c \in \mathbf{K}$. Then, $x - \beta$ is a uniformization parameter. Since $dx = d(x - \beta)$, then there is neither a zero nor a pole.

(B) Suppose that $c \notin \mathbf{K}$. Then, $x - \beta$ is a uniformization parameter. Since $dx = d(x - \beta)$, then there is neither zero nor pole.

(3) Note that with our basic assumption that the field $\mathbf{K}$ is perfect, $f(x), f'(x)$ have no common root. Clearly, we have $df(x) = f'(x)dx$ or $dx = (\frac{1}{f'(x)})df(x)$. It can be shown that $f(x)$ is a local uniformization parameter in $\mathbf{K}[x]_{(f(x))}$ and at the corresponding point(s) on the curve C. Since the residue of $(\frac{1}{f'(x)})$ is not zero modulo $(f(x))$, then there is neither a zero nor a pole.

Let us consider the point at ∞. As pointed out in **Example 19** in Section 4.6, $x = t^{-2}$ for a uniformization parameter t. Therefore, $dx = -2t^{-3}dt$, i.e., it has three poles at ∞. We conclude that $(dx) = \mathbf{P}_0 + \mathbf{P}_1 + \mathbf{P}_a - 3\mathbf{P}_\infty$ and degree of $((dx)) = 0 = 2g - 2$ and $g = 1$. ∎

Proposition 4.44. *We always have*

$$i(\mathbf{D}) = \ell(\mathbf{W} - \mathbf{D}),$$

where $\mathbf{W}$ is a canonical divisor.

Proof. Let $\mathbf{W}=(fdx)$. We have

$$g \in L(\mathbf{W} - \mathbf{D}) \Leftrightarrow$$
$$(g) + (\mathbf{W} - \mathbf{D}) \succ 0 \Leftrightarrow$$
$$(g) + (\mathbf{W}) \succ \mathbf{D} \Leftrightarrow$$
$$(g) + (fdx) \succ \mathbf{D} \Leftrightarrow$$
$$gfdx \in \Omega(\mathbf{D}).$$

Therefore, there is a bijective map from $L(\mathbf{W} - \mathbf{D})$ to $\Omega(\mathbf{D})$. They must be of the same dimension. Our proposition follows from the definitions of $\ell(\mathbf{W} - \mathbf{D})$ (cf. Definition 4.23) and of $i(\mathbf{D})$ (cf. Definition 4.33). □

Example 29: Let us use the notations of **Example 28**. Let $\mathbf{D} = \mathbf{P}_0 + \mathbf{P}_1 + \mathbf{P}_a - 3\mathbf{P}_\infty$. Certainly, we know that $(dx) \succ \mathbf{D}$. Let fdx be another differential such that $(fdx) \succ \mathbf{D}$. Then clearly, we have $(f) \succ \mathbf{0}$. Therefore, $f = const$. Hence, $i(\mathbf{D}) = 1$. Let us consider another example. Let $\mathbf{D}' = n\mathbf{P}_\infty$, where n is a positive integer. Note that $d(\mathbf{W} - \mathbf{D}') < 0$. Therefore, $0 = \ell(\mathbf{W} - \mathbf{D}') = i(\mathbf{D}')$. ∎

Proposition 4.45. *If $d(\mathbf{D}) > 2g - 2$, then $i(\mathbf{D}) = 0$.*

Proof. Clearly, $d(\mathbf{W} - \mathbf{D}) < 0$, and $0 = \ell(\mathbf{W} - \mathbf{D}) = i(\mathbf{D})$. □

In the previous discussions, sometimes we extend the ground field from $\mathbf{K}$ to a separable extension $\mathbf{K}'$. We have the following proposition.

Proposition 4.46. *If $\mathbf{K}'$ is a separable extension of $\mathbf{K}$ (which is always true for $\mathbf{K}$ a perfect field), then there is no change of genus g for the ground field extension.*

Proof. Chevalley, p. 99 [10]. $\square$

The above proposition tells us that when we compute the *genus*, we may extend the ground field to the algebraic closure of $\mathbf{K}$ and use the corresponding result about *genus* from *algebraic geometry*, say, using the Proposition 4.30 (Plücker's formula).

Example 30: Let us consider **Example 19** of the preceding section. We have an equation which defines a curve C of genus 1 with $a \neq 0, 1$:

$$y^2 = (x - a)(x - 1)(x - 0).$$

Let us use the notations of **Example 29**. According to the Riemann–Roch theorem, we have

$$\ell(\mathbf{D}) - i(\mathbf{D}) = 0.$$

Since $i(\mathbf{D}) = 1$, so $\ell(\mathbf{D}) = 1$. A generator of $L(\mathbf{D})$ is $\frac{1}{y}$. Let us consider $\mathbf{D}' = n\mathbf{P}_\infty$. Then, it follows from the Riemann–Roch theorem that $\ell(\mathbf{D}') = n$ since we have $i(\mathbf{D}') = 0$ and $g = 1$. Clearly, the set $\{x^k y^j : j = 0, 1, \text{and } 2k + 3j \leq n\}$ forms a basis for $L(\mathbf{D})$.   ∎

Exercises

(1) Given a smooth and regular curve C, the Weierstrass gaps is the integers i such that there is no rational function z which has a pole only at $\mathbf{P}$ and $\mathrm{ord}_{\mathbf{P}}(z) = -i$. Show that there are precisely g Weierstrass gaps where g is the genus of C.

(2) For **Example 29**, show that $\ell(\mathbf{D}) = 1$ directly.

(3) Let $\mathbf{P}$ be a rational point on P_K^1. Show that $i(\mathbf{P}) = 0$ and $\ell(\mathbf{P}) = 2$.

(4) Given the curve $C = P_K^1$, find a differential $f(x)dx$ with residues 1 at $x = 1$ and -1 at $x = -1$ with 0 residues elsewhere.

(5) Given the curve $P_{\mathbb{C}}^1$, where $\mathbb{C}$ is the field of complex numbers and any divisor $\mathbf{D}$, show that the dimension of all linear functional on Ξ which vanishes on $\Xi(-\mathbf{D}) + F$ is $\max(0, d(\mathbf{D})\text{-}1)$.

4.9. Weil's Conjecture and Hasse–Weil Inequality

Given an smooth projective algebraic curve C in *algebraic geometry coding theory*, it is important to know the number of rational points on the curve over a finite field F_q. Let us extend the ground field $\mathbf{F}_q$ to $\mathbf{F}_{q^r}$. Let N_r be the number of rational points of $\bar{C}$ over $\mathbf{F}_{q^r}$. The **Weil's conjecture** is a conjecture about the *zeta function* of $\mathbf{C}$ which is related to N_r.

Definition 4.47. The *zeta function* $\mathcal{Z}(t)$ is defined as

$$\mathcal{Z}(t) = \exp\left(\sum_{r=1}^{\infty} N_r \frac{t^r}{r}\right). \qquad \blacksquare$$

Clearly, we know the function $\mathcal{Z}(t)$ iff we know all N_r. We shall try to locate $\mathcal{Z}(t)$.

Proposition 4.48 (Weil's conjecture for curves). *We have for a smooth projective curve over a finite field $\mathbf{F}_q$ of genus g the following:*

$$\mathcal{Z}(t) = \frac{P_1(t)}{(1-t)(1-qt)},$$

where

$$P_1(t) = \prod_{i=1}^{2g}(1 - \alpha_i t),$$

with $\mid \alpha_i \mid = \sqrt{q}$.

Proof. The conjecture was proved by Weil, see [40]. $\qquad \square$

Remark: The Weil's conjecture for higher dimensional projective algebraic varieties is established by Deligne (1974). $\qquad \blacksquare$

Example 31: Let us consider $C = P^1_{q^r}$. It is easy to see that C has q^r rational points on $A^1_{q^r}$ and one point extra (the point at ∞). So, there are $q^r + 1$ points and $N_r = q^r + 1$. Substituting it in the definition of $\mathcal{Z}(t)$, we have

$$\mathcal{Z}(t) = \exp\left(\sum_{1}^{\infty}(q^r + 1)\left(\frac{t^r}{r}\right)\right)$$

$$= \exp\left(\sum_{1}^{\infty}(q^r)\left(\frac{t^r}{r}\right)\right)\exp\left(\sum_{1}^{\infty}(1)\left(\frac{t^r}{r}\right)\right).$$

Recall that

$$\ln\left(\frac{1}{1-x}\right) = -\sum_{1}^{\infty}\left(\frac{x^r}{r}\right).$$

Therefore, we deduce that

$$\mathcal{Z}(t) = \frac{1}{(1-t)(1-qt)}.$$ ∎

Proposition 4.49 (Hasse–Weil's inequality). *Let C be a smooth and absolute irreducible projective curve. Then, we have*

$$\mid N_1 - q - 1\mid \le 2g\sqrt{q}.$$

Proof. We may deduce a proof in the curves case from Hartshorne's book "Algebraic Geometry" V Ex 1.10, Appendix C Ex 5.7.   □

Example 32: Let $p \ge 5$ and $q = p^m$. Then, the Fermat curve $x_0^{q-1} + x_1^{q-1} + x_2^{q-1}$ has no solution as a projective curve. Therefore, $N_1 = 0$, and $g = \frac{(q-2)(q-3)}{2}$. The Hasse–Weil's inequality is to show that $(q-2)(q-3)\sqrt{q} \ge q + 1$. Let $f(q) = (q-2)(q-3)\sqrt{q} - q - 1$. It is easy to see that $f(5) > 0$ and $f'(x) > 0$ for all $x \ge 5$. It is easy to see $f(q) > 0$ for all $p \ge 5$ and Hasse–Weil's inequality is satisfied in those cases.   ∎

Example 33: Let us consider the smooth curve C defined by $x_0^3 + x_1^3 + x_2^3$ over $\mathbf{F}_q$. It is easy to see that genus $g = 1$. Let us take $q = 2$. Let us count the number of rational points on C. Since $x_i^3 = x_i$, it is easy to see that two of x_i's must be 1 and the third is 0. So, we have three rational points.

Let $q = 2^2$. Let us count the number of rational points on C. Note that $\alpha^3 = 1$ for any non-zero $\alpha \in \mathbf{F}_{2^2}$. It is impossible to have two of x_0, x_1, x_2 to be 0 since the third one must be 0 and $(0,0,0)$ is not a projective point. It is also impossible to have x_0, x_1, x_2 all non-zero. Note that then $x_i^3 = 1$ for all i, we have the impossible situation of $1 + 1 + 1 \ne 0$. Therefore, we must have exactly one zero. Note that if several $x_i \ne 0$, then we may assume that one of $x_i = 1$. Then, we have the following:

(1) $x_0 = 0, x_1 = 1$, x_2 arbitrarily non-zero $\in \mathbf{F}_{2^2}$. There are three points.
(2) $x_1 = 0, x_2 = 1$, x_0 arbitrarily non-zero $\in \mathbf{F}_{2^2}$. There are three points.
(3) $x_2 = 0, x_0 = 1$, x_1 arbitrarily non-zero $\in \mathbf{F}_{2^2}$. There are three points.

So, there are totally nine points. This makes the non-restrictive inequality in Hasse–Weil's inequality an equality.

Let $q = 2^3$. We shall use brute force to compute the number of rational points. It is impossible to have two of x_0, x_1, x_2 to be 0 since the third one must be 0 and $(0,0,0)$ is not a projective point. So, we have the following two cases: (1) two of them are non-zero and the third zero; (2) all of them are non-zero.

Case (1) Let us consider $x_2 = 0$, $x_1 = 1$. Then, we have

$$x_0^3 = 1.$$

The above equation is satisfied only if $x_0 = 1$. Therefore, by taking $x_0 = 0$ or $x_1 = 0$, we conclude that there are three rational points.

Case (2) We may take $x_2 = 1$. Let α be a generator of the multiplicative group $F_{2^3}^*$. Then, $x^7 = 1$ for any non-zero x. If $x_1^3 = 1$, then $x_1 = 1$, and we must have $x_0 = 0$. This point has been counted. Therefore, we have that $x_1 \neq 1$, or $x_1 = \alpha, \alpha^2, \ldots, \alpha^6$. We want to show that for every possible value of x_1, we may find one and only one value for x_0 to satisfy the equation. Let us consider $x_1 = \alpha^j$ with $1 \leq j \leq 6$. Let x_0 $\bar{x}_0$ be two solutions with fixed $x_1 = \alpha^j$. Then, we have

$$1 + \alpha^{3j} = \alpha^k = \alpha^{k+7} = \alpha^{k+14} = x_0^3 = \bar{x}_0^3.$$

Then, it is easy to see that

$$\left(\frac{\bar{x}_0}{x_0}\right)^3 = 1.$$

Therefore, $\bar{x}_0 = x_0$, so the solution must be unique. Moreover, 3 is a factor of one of $k, k+7, k+14$; therefore, the equation can be solved.

So, there are totally nine rational points.

Let $q = 2^4$. We shall use brute force to compute the number of rational points. Let α, β be defined by the following equations:

$$\alpha^2 + \alpha + 1 = 0,$$

$$\beta^2 + \beta + \alpha = 0.$$

It is easy to see that $\mathbf{F}_{2^2} = \mathbf{F}_2[\alpha]$ and $\mathbf{F}_{2^4} = \mathbf{F}_{2^2}[\beta]$. Let us count the number of rational points on C. It is impossible to have two of x_0, x_1, x_2 to be 0 since the third one must be 0 and $(0,0,0)$ is not a projective point. So, we have the following two cases: (1) two of them are non-zero and the third zero; (2) all of them are non-zero.

Case (1) Let us consider $x_2 = 0$, $x_1 = 1$. Then, we have

$$x_0^3 = 1.$$

The above equation is satisfied only if $x_0 \in \mathbf{F}_{2^2} \subset \mathbf{F}_{2^4}$. There are 3 possible points. Therefore, by taking $x_0 = 0$ or $x_1 = 0$, we conclude that there are 9 rational points.

Case (2) We may take $x_2 = 1$. Then, the defining equation can be rewritten as

$$x_0^3 + 1 = x_1^3.$$

Let $y = x_0^3 \neq 0$. Since $x_1 \neq 0$, we have

$$y^5 = x_0^{15} = 1,$$
$$(y + 1)^5 = x_1^{15} = 1.$$

It is easy to see that the above equations can be rewritten as

$$y^4 + y^3 + y^2 + y + 1 = 0,$$
$$y^4 + y + 1 = 0.$$

Then, $y = 0$ or 1 which are impossible. Therefore, there is no rational point in this case. Totally, we have 9 rational points.

We may use the *Weil's Conjecture* for the same purpose to determine the numbers of points for all F_{2^r} once we know the number $N_1 = 3$ and $N_2 = 9$. Let us compute N_3. Note that

$$\mathcal{Z}(t) = \exp\left(\sum_{r=1}^{\infty} N_r \frac{t^r}{r}\right),$$

$$\mathcal{Z}(t) = \frac{P_1(t)}{(1 - t)(1 - qt)},$$

where

$$P_1(t) = \prod_{i=1}^{2g}(1 - \alpha_i t) = (1 - \alpha t)(1 - \beta t)$$

since $g = 1$. Then, we have

$$ln\ \mathcal{Z}(t) = 3t + 9t^2/2 + N_3 t^3/3 + \cdots = ln\ (1 - \alpha t) + ln\ (1 - \beta t)$$
$$- ln\ (1 - t) - ln\ (1 - 2t) = (1 + 2 + \alpha + \beta)t$$
$$+ (1 + 2^2 + \alpha^2 + \beta^2)/2t^2 + (1 + 2^3 + \alpha^3 + \beta^3)/3t^3 + \cdots.$$

Comparing coefficients of the powers of t, we have the following equations:

$$(1)\ 1 + 2 + \alpha + \beta = 3,$$
$$(2)\ 1 + 4 - \alpha^2 - \beta^2 = 9,$$
$$(3)\ 1 + 2^3 - \alpha^3 + \beta^3 = N_3.$$

Solving the first two equations, we have

$$\alpha = \sqrt{-2}, \beta = -\sqrt{-2}.$$

Substituting into equation (3), we conclude that $N_3 = 9$. It checks with our previous computation.

Note that *Hasse* proved the following formula for any genus 1 smooth projective curve:

$$N_r = 1 + q^r - \alpha^r - \beta^r.$$

Our result checks with the formula. ∎

Example 34: We want to show that the *zeta function* is not determined by the genus of the curve. Let us consider the following genus one curve:

$$y^2 = x(x - 1)(x - \alpha)$$

over F_{2^2}, with $\alpha \neq 0, 1$ in F_{2^2}. Then, F_{2^2} consists of $0, 1, \alpha, \alpha + 1$ and α satisfies the following equation:

$$\alpha^2 + \alpha + 1 = 0.$$

Let $x = 0, 1, \alpha$, then $y = 0$. Let $x = \alpha + 1$, then $y = 1$, we have another point at ∞; therefore, there are five rational points. In the previous example, there are nine points over F_{2^2}. Therefore, the *zeta functions* must be different. ∎

Example 35: Let us consider a curve C defined by $x_0^5 + x_1^5 + x_2^5 = 0$ over F_{2^4}. Let us count the number of rational points on C. It is impossible to have two of x_0, x_1, x_2 to be 0 since the third one must be 0 and $(0, 0, 0)$ is not a projective point. Let us consider the case that there is exactly one zero. Then, we have

(1) $x_0 = 0, x_1 = 1, x_2^5 = 1$. Note that $x^5 + 1 \mid x^{15} + 1$ by the field equation for F_{2^4}, there are 5 distinct x_2 satisfying $x_2^5 = 1$. There are five points.
(2) $x_1 = 0, x_0 = 1$, similarly there are five points.
(3) $x_2 = 0, x_1 = 1$, similarly there are five points.

Suppose that there is no zero among x_0, x_1, x_2. We may take $x_0 = 1$. Let $x_1^5 = a \neq 0$, then $x_2^5 = 1 + a \neq 0$. It means that $a \neq 0, 1$. Since $(x_1^5)^3 = 1 = (x_2^5)^3$, we must have $a^3 = (a+1)^3 = 1$. Or $a, a+1 \in \mathbf{F}_{2^2}$, Let $\mathbf{F}_{2^2} = \mathbf{F}_2[\alpha]$. Then, $(a, a+1) = (\alpha, \alpha+1)$ or $(\alpha+1, \alpha)$. For the equations $x_1^5 = a$, $x_2^5 = 1 + a$, there are five distinct solutions for each, so there are 25 solutions for one pair $(a, a+1)$. There are 50 solutions. The total number of solutions is 65. Note that $g = 6$ and $\mid 65 - 6 - 1 \mid \leq 2 \times 6 \times 8$. The Hasse–Weil's inequality is satisfied.   ∎

Example 36: For the later applications in coding theory, let us consider the Klein quartics curve over $\mathbf{F}_2$ or $\mathbf{F}_{2^2}$ or $\mathbf{F}_{2^4}$ defined by the equation $x^3 y + y^3 + x = 0$. Its genus is $g = 3$.

For coding theory, let us count the number of rational points. It is clear that $x = 0 \Leftrightarrow y = 0$. We denote this point $(0,0)$ (or $(1,0,0)$ as the projective point) by $\mathbf{P}$. It is easy to see that $(1,1)$ is not a point. After we homogenize the equation, it becomes $x^3 y + y^3 z + xz^3 = 0$. Let $z = 0$. Then, the curve has other two points at ∞, $(0,0,1) = \mathbf{P}_1$ and $(0,1,0) = \mathbf{P}_2$ over any field $\mathbf{K}$. So, there are three rational points over $\mathbf{F}_2$:

$$\mathbf{P} = (1,0,0), \quad \mathbf{P}_1 = (0,0,1), \quad \mathbf{P}_2 = (0,1,0).$$

Now, let us count the points at the finite distance. Over $\mathbf{F}_{2^2}$, we have $x^3 = 1$ for all $x \neq 0$. Therefore, the affine equation gets reduced to $y + 1 + x = 0$ if $x, y \neq 0$, i.e., $y = 1 + x$, where $x \neq 1, 0$. Let α be a field generator of $\mathbf{F}_{2^2}$ over $\mathbf{F}_2$, i.e., $\alpha^2 + \alpha + 1 = 0$. Then, $(\alpha, \alpha+1) = \mathbf{P}_3, (\alpha+1, \alpha) = \mathbf{P}_4$ are the extra rational points on the curve. Therefore, over $\mathbf{F}_{2^2}$, there are five rational points. The two extra points are

$$\mathbf{P}_3 = (\alpha, \alpha+1), \quad \mathbf{P}_4 = (\alpha+1, \alpha).$$

Let us consider the field $\mathbf{F}_{2^4}$. Let $\mathbf{F}_{2^4} = \mathbf{F}_{2^2}[\beta]$ with β satisfying $\beta^2 + \beta + \alpha = 0$. First, we consider $x = \alpha$. Note that $\alpha^3 = 1$. Therefore, the equation will be reduced to $y^3 + y + \alpha = 0$. It is easy to see that $y^3 + y + \alpha = (y + \alpha + 1)(y^2 + (\alpha+1)y + (\alpha+1))$. We have to solve $y^2 + (\alpha+1)y + (\alpha + 1) = 0$. The two roots are $y = \alpha^2(\beta+1), \alpha^2\beta$. We name $(\alpha, \alpha^2(\beta)) = \mathbf{P}_5$, $(\alpha, \alpha^2(\beta+1)) = \mathbf{P}_6$. Similarly, we find $\mathbf{P}_7 = (\alpha+1, (\alpha+1)^2(\beta+\alpha))$, $\mathbf{P}_8 = (\alpha+1, (\alpha+1)^2(\beta+\alpha+1))$. Thus, we have four more points:

$$\mathbf{P}_5 = (\alpha, \alpha\beta + \beta), \quad \mathbf{P}_6 = (\alpha, \alpha\beta + \beta + \alpha + 1),$$
$$\mathbf{P}_7 = (\alpha+1, \alpha\beta + +\alpha + 1), \quad \mathbf{P}_8 = (\alpha+1, \alpha\beta + \alpha).$$

We may consider the possibility that $y = \alpha, \alpha + 1$. Let $y = \alpha$. Then, the equation will reduce to $\alpha x^3 + x + 1 = 0$. Replacing x by αu, and the equation becomes $u^3 + u + \alpha + 1 = 0$, then the equation is similar to the one we just discussed. We conclude that there are four more points:

$$\mathbf{P}_9 = (\alpha\beta + \beta + 1, \alpha), \quad \mathbf{P}_{10} = (\alpha\beta + \beta + \alpha, \alpha),$$

$$\mathbf{P}_{11} = (\alpha\beta, \alpha + 1), \quad \mathbf{P}_{12} = (\alpha\beta + \alpha, \alpha + 1).$$

Let us consider $y = \alpha\beta$. Then, $\mathbf{P}_{13} = (\beta, \alpha\beta)$. Let us consider $y = \alpha(\beta + 1)$. Then, $\mathbf{P}_{14} = (\beta + 1, \alpha\beta + \alpha)$. Let us consider $y = \alpha\beta + \beta + 1$. Then, $\mathbf{P}_{15} = (\alpha + \beta, \alpha\beta + \beta + 1)$. Let us consider $y = \alpha\beta + \beta + \alpha$. Then, $\mathbf{P}_{16} = (\alpha + \beta + 1, \alpha\beta + \beta + \alpha)$.

There are 17 rational points over $\mathbf{F}_{2^4}$. ∎

Exercises

(1) Count the number of solutions of $x^2 + y^2 = 1$ over $\mathbb{Z}_2$. Do the same over $\mathbb{Z}_5$.

(2) Do not use the propositions of this section. Show that a projective curve of genus 0 over $\mathbf{F}_q$ has at most $q + 1$ rational smooth points.

(3) Let $\{C_s\}$ be a sequence of curves with genus g_s and the number of rational smooth point N_s. Show that if $N_s \mapsto \infty$, then $g_s \mapsto \infty$.

(4) Show that the numbers of rational points for the projective curve in **Example 33** satisfy the Hasse–Weil's inequality.

(5) Use **Example 33** and Weil's conjecture to compute Weil's *zeta function* for that particular projective curve.

(6) What is the number of rational points for the projective curve in **Example 33** over the field $\mathbf{F}_{2^5}$?

(7) What is the number of rational points for the projective curve in **Example 33** over the field $\mathbf{F}_{2^6}$?

(8) Find the *zeta function* of the curve in **Example 34**.

PART IV
Algebraic Geometric Codes

Chapter 5

Algebraic Curve Goppa Codes

5.1. Geometric Goppa Codes

We use the term *the theory of smooth algebraic projective curves over a finite field $\mathbf{F}_q$* for *the theory of functions of one variable over a finite field F_q*.

As we all know that after Shannon, the theory of coding theory were separated into two parts: (1) theoretical part about the existence of good codes, (2) decoding procedures. Note that for Hamming code, or more general for any vector-space code, we have the following diagram:

$$\text{message space } \mathbf{F}_q^k \xrightarrow{\pi} \text{word space } \mathbf{F}_q^n,$$

where the map π is an injective map with the image of the code space. For the later developments, we slightly modify the coding theory to the following diagram:

$$\text{message space } \mathbf{F}_q^k \xrightarrow{\sigma_1} \text{function space } \xrightarrow{\sigma_2} \text{word space } \mathbf{F}_q^n.$$

The first map σ_1 is injective. Thus, we use functions to rename the messages, and the second map σ_2 is a homomorphism with the image of $\sigma_2\sigma_1(\mathbf{F}_q^k) =$ as the code space. Usually, the map σ_2 is an evaluation map which evaluates a function f at an ordered n-tuple of points $(P_1, \ldots, P_n)$. Thus, it maps a function f to $[f(P_1), \ldots, f(P_n)] \in \mathbf{F}_q^n$. Note that $\sigma_2\sigma_1$ will send the message space to the word space, certainly we do not want to send any non-zero message to zero; thus, we require that the composition $\sigma_2\sigma_1$ is an injective map on the message space. In our previous discussions, the message space is naturally mapped to the code space. The theorists are mainly working on the function space, and the engineers work on the methods to correct the errors after the transmissions of the code words. For Reed–Solomon codes, the function space is all polynomials with degree $k - 1$ or less, which is a subspace of all polynomials with degree $n - 1$ or less (these pair of vector spaces are mapped to a pair of a code space $\subset$ the word space $= \mathbf{F}_q^n$ by evaluations at a sequence of points). For classical Goppa codes, the function space is the set of rational functions of the following form:

$$f = \sum_{i=1}^{n} \frac{c_i}{(x - \gamma_i)} \equiv 0 \quad \mathrm{mod} \ (g(x))$$

and $\sigma_2(f)$ is the n-tuples $[c_1, \ldots, c_n]$.

In 1981, Goppa discovered an amazing connection between the theory of algebraic smooth curves over a finite field $\mathbf{F}_q$ and the theory of error-correcting block q-ary codes. He allowed the function space to be a subspace of $L(\mathbf{D})$ for some divisors $\mathbf{D}$ on the smooth curve. The idea is quite simple and generalizes the well-known construction of Reed–Solomon codes and the classical Goppa codes. The Reed–Solomon codes use polynomials in one variable over $\mathbf{F}_q$ (the rational functions over $\mathbf{P}_{\mathbf{F}_q}^1$ with only pole at ∞) (see Example 1), and the classical Goppa codes use the *residue form* over $P_{\mathbf{F}_q}^1$ (see Example 3). Goppa generalized those ideas using rational functions or differentials on an algebraic projective curve (these two versions are equivalent, see later discussions). In coding theory, we have message space $\mathbf{F}_q^k$, the k-dimensional code space, and the word space $U = \mathbf{F}_q^n$. Taking any basis of the k-dimensional subspace of the function space, we may specify a basis $\{\phi_i\}$ for the subspace and a map $\pi([a_1, \ldots, a_k]) = \sum_i a_i \phi_i$ from the message space to the function space, and further, define the second map which complete the coding process. In terms of linear algebra, what we require is

$$image(\sigma_1) \cap ker(\sigma_2) = \{0\}.$$

Note that if the dimension of $image(\sigma_1)$ is k, then the *rate of information* is k/n (cf. Definition 1.27). Certainly, we want to consider the maximally possible k, i.e., **we shall thus maximize the message space by requiring in the rest of the book that**

$$image(\sigma_1) \oplus ker(\sigma_2) = \text{the function space.}$$

This part of transforming messages to function space is easy. We shall discuss the map σ_2 of the function space to $\mathbf{F}_q^n$ first.

Definition 5.1 (Algebraic geometric code or a geometric Goppa code in function form $C_L(\mathbf{B}, \mathbf{D})$). Let $\mathbf{C}$ be an absolutely irreducible smooth projective algebraic curve of genus g over $\mathbf{F}_q$. Consider an (ordered) set $\{\mathbf{P}_1, \mathbf{P}_2, \ldots, \mathbf{P}_n\}$ of distinct $\mathbf{F}_q$ rational points on $\mathbf{C}$ and a divisor $\mathbf{B} = \sum_{i=1}^{n} \mathbf{P}_i$ and a divisor $\mathbf{D}$ on $\mathbf{C}$. For simplicity, let us assume that the support of $\mathbf{D}$ is disjoint from the support of $\mathbf{B}$; therefore, $f(\mathbf{P}_j) \neq \infty$ for $f \in L(\mathbf{D})$ for all j (i.e., $L(\mathbf{D})$ is the function space). The linear evaluation map Ev_B (i.e., σ_2) will send the linear space $L(\mathbf{D})$ of rational functions on C (associated to $\mathbf{D}$) to the word space $\mathbf{F}_q^n$.

$$Ev_B : L(\mathbf{D}) \mapsto \mathbf{F}_q^n.$$

$$f \mapsto (f(\mathbf{P}_1), \ldots, f(\mathbf{P}_n)). \qquad \blacksquare$$

Remark 1: If we take $\mathbf{D}$ as $m\mathbf{P}$ for some rational point $\mathbf{P}$, then it is called a *one-point code*. $\qquad \blacksquare$

It is easy to see that the word space is $\mathbf{F}_q^n$ in n-dimensional vector space. We wish to find the parameters $[n, k, d]$ for this code. Clearly, $n = \dim(\mathbf{F}_q^n)$. A function $f \in L(\mathbf{D})$ is sent to $[0 \cdots 0] \Leftrightarrow f(\mathbf{P}_j) = 0$ for all j, i.e., $f \in L(-\mathbf{B}) \Leftrightarrow$ (since the supports of $\mathbf{B}$ and $\mathbf{D}$ are disjoint) $f \in L(\mathbf{D} - \mathbf{B})$. Therefore, we have the code space which is canonically isomorphic to $L(\mathbf{D})/L(\mathbf{D} - \mathbf{B})$; it means that with $V = image(\sigma_1)$, we have

$$V \oplus L(\mathbf{D} - \mathbf{B}) = L(\mathbf{D}),$$

where V $(=\sigma_1(message\ space))$ is isomorphic to the code space. Let $k =$ dimension of $V =$ dimension of code space, then we have

$$k = \ell(\mathbf{D}) - \ell(\mathbf{D} - \mathbf{B}).$$

Once we have the function space, let $\{\phi_1, \ldots, \phi_k\}$ be the basis for the subspace V, then a map $\pi([a_1, \ldots, a_k]) = \sum_i a_i \phi_i$ will define a map from the message space to $V \subset$ the function space.

The important cases for applications are $0 < d(\mathbf{D}) < n$; we find the indices k and d. We have the following proposition.

Proposition 5.2. *Let us use the notations of Definition 5.1. Let us assume that $0 < d(\mathbf{D}) < n$ and $d(\mathbf{D}) > g - 1$, then $\ell(\mathbf{D} - \mathbf{B}) = 0$ and the code space is isomorphic to $L(\mathbf{D})$, and*

$$k = \ell(\mathbf{D}) \geq d(\mathbf{D}) - g + 1 > 0.$$

The minimal distance d satisfies

$$d \geq n - d(\mathbf{D}).$$

We have the following:

$$n + 1 \geq k + d \geq n + 1 - g.$$

Proof. In this case, we have $d(\mathbf{D} - \mathbf{B}) < 0$ and $\ell(\mathbf{D} - \mathbf{B}) = 0$. Therefore, $L(\mathbf{D} - \mathbf{B}) = \{0\}$, and the code space $V = L(\mathbf{D})$. In view of Riemann's theorem, we have

$$k = \ell(\mathbf{D}) \geq d(\mathbf{D}) - g + 1 > 0.$$

We prove the inequality $d \geq n - d(\mathbf{D})$. Let $\mathbf{D}$ be written as the difference of two disjoint effective divisors $\mathbf{D} = \mathbf{D_0} - \mathbf{D_\infty}$. Since $k > 0$, we may pick $0 \neq f \in L(\mathbf{D})$. Let $\mathbf{B}'$ be $\sum \mathbf{P}_j$, where $\mathbf{B} \succ \mathbf{P}_j$ and $f(\mathbf{P}_j) = 0$. Clearly, we have $\mathbf{B} \succ \mathbf{B}'$. Let us separate the proof into the following two cases: either (1) $d(\mathbf{B}') \leq d(\mathbf{D})$ or (2) $d(\mathbf{B}') > d(\mathbf{D})$. (1) Note that if $d(\mathbf{B}') \leq d(\mathbf{D})$, then the distance $d(f, 0) = n - d(\mathbf{B}') \geq n - d(\mathbf{D})$, recall that $d = \min\{d(f, 0) : f \in L(\mathbf{D})\}$, which is the inequality $d \geq n - d(\mathbf{D})$ we wish to prove.

The relation $(f) + \mathbf{D} = (f)_0 - (f)_\infty + \mathbf{D_0} - \mathbf{D_\infty} \succ 0$ means that $(f)_0 \succ \mathbf{D_\infty}$, $\mathbf{D_0} \succ (f)_\infty$ and $(f)_0 \succ \mathbf{B}'$. Since $\mathbf{B}$ and $\mathbf{D}$ are disjoint, $\mathbf{B}'$, $\mathbf{D_\infty}$ are disjoint, then it is easy to conclude that $(f)_0 \succ \mathbf{D_\infty} + \mathbf{B}'$. Let us consider the case (2), i.e., $d(\mathbf{B}') > d(\mathbf{D})$. Let $d(\mathbf{D}) = d(\mathbf{D_0}) - d(\mathbf{D_\infty}) = r - s$, where $r = d(\mathbf{D_0})$ and $s = d(\mathbf{D_\infty})$. Then, $d((f)_0) \geq d(\mathbf{D_\infty}) + d(\mathbf{B}') > s + d(D) = s + r - s = r \geq d((f)_\infty)$. Therefore, $f = 0$. A contradiction to our assumption that $f \neq 0$. We prove that

$$d \geq n - d(\mathbf{D}).$$

For the last inequalities, the first one is the classical Singleton bound (Proposition 1.23). The second one comes from the two inequalities already established for k and d of this proposition as

$$k + d \geq d(\mathbf{D}) - g + 1 + d \geq n + 1 - g. \qquad \square$$

Remark 2: The number k provides the *rate of information* k/n, and the lower bound $n - d(\mathbf{D})$ for d provides the bound number $\lfloor \frac{(n - d(\mathbf{D}))}{2} \rfloor$ for the number t of corrections for the code (cf. the remark after Proposition 1.25, as long as $t \leq \lfloor \frac{(n - d(\mathbf{D}) - 1)}{2} \rfloor$, the code can correct t errors by brute force). The lower bounds for the rank $(d(\mathbf{D}) - g + 1)$ and the minimal distance $(n - d(\mathbf{D}))$ are called the **designed rank** and **designed minimal distance**. ∎

Example 1: Let the curve C be the projective line $P^1_{F_q}$ over a finite field $\mathbf{F}_q$. Let $n = q - 1$ and β be a generator of the cyclic group $\mathbf{F}^*_q$. Let $\mathbf{B} = \mathbf{P}_\beta + \mathbf{P}_{\beta^2} + \cdots + \mathbf{P}_{\beta^n}$, where $\mathbf{P}_{\beta^j}$ is the point that solves the equation $x - \beta^j = 0$, and $\mathbf{D} = (k - 1)\mathbf{P}_\infty$, where $\mathbf{P}_\infty$ is the point at infinity, and we assume that $k < n$. Then, we have

$$L(\mathbf{D}) = \text{the set of all polynomials of degree less than k over } \mathbf{F}_q.$$

It is then easy to see that this is the classical **Reed–Solomon code**. ∎

Example 2: Let us consider curve C defined by $x_0^3 + x_1^3 + x_2^3$ over $\mathbf{F}_{2^2}$. It is a curve of genus 1. According to Example 33 of Chapter 4, there are nine rational points. We may take eight of them and call them $\mathbf{P}_1, \mathbf{P}_2, \mathbf{P}_3, \mathbf{P}_4, \mathbf{P}_5, \mathbf{P}_6, \mathbf{P}_7, \mathbf{P}_8$, and call the other one $\mathbf{P}$. Let $\mathbf{B} = \sum_{i=1}^8 \mathbf{P}_i$ and $\mathbf{D} = m\mathbf{P}$. Then, it follows from the preceding proposition that

$$k \geq m - 1 + 1 = m,$$

$$d \geq 8 - m.$$

If we select $m = 3$, then $k \geq 3, d \geq 5$ and the code will have at least a rate of information $\frac{3}{8}$, and later on, we shall show that it corrects at least $\lfloor \frac{d-1}{2} \rfloor$ ≥ 2 errors. It is distinct from the Hamming code which corrects only 1 error (cf. the **remark** after Proposition 1.25). ∎

Originally, Goppa used the differentials on $\mathbf{C}$ to form the dual construction of the above as follows.

Definition 5.3 (Algebraic geometric code or a geometric Goppa code in residue form $C_\Omega(\mathbf{B}, \mathbf{D})$). Let C be an absolutely irreducible smooth projective algebraic curve of genus g over $\mathbf{F}_q$. Consider an (ordered) set $\{\mathbf{P}_1, \mathbf{P}_2, \ldots, \mathbf{P}_n\}$ of distinct $\mathbf{F}_q$ rational points on C and a divisor $\mathbf{B} = \sum_{i=1}^n \mathbf{P}_i$, and a $\mathbf{F}_q$-divisor $\mathbf{D}$ on C. For simplicity, we assume that the support of $\mathbf{D}$ is disjoint from the support of $\mathbf{B}$. Let the space of differentials be $\Omega(\mathbf{D} - \mathbf{B})$ and the map σ_2 be defined as the following map res_B.

$$\text{res}_B : \Omega(\mathbf{D} - \mathbf{B}) \mapsto \mathbf{F}_q^n,$$

$$\eta \mapsto (\text{res}_{\mathbf{P}_1} \eta, \ldots, \text{res}_{\mathbf{P}_n} \eta). \qquad ∎$$

We wish to find the parameters $[n, k, d]$ for this code. Clearly, $n = \dim(\mathbf{F}_q^n)$. Note that $\Omega(\mathbf{D}) \subset \Omega(\mathbf{D} - \mathbf{B})$. A differential $\eta \in \Omega(\mathbf{D} - \mathbf{B})$ is sent to $[0 \cdots 0] \Leftrightarrow \mathrm{res}_{\mathbf{P}_j} \eta = 0$ for all j, i.e., $(\eta) \succ \mathbf{D} \Leftrightarrow \eta \in \Omega(\mathbf{D})$. There is a vector subspace $V \subset \Omega(\mathbf{D} - \mathbf{B})$ with $V \oplus \Omega(\mathbf{D}) = \Omega(\mathbf{D} - \mathbf{B})$, and

$$k = i(\mathbf{D} - \mathbf{B}) - i(\mathbf{D}).$$

Proposition 5.4. *A geometric Goppa code in residue form defines an* $[n, k, d]$ *code in the same way as above; then, we have* $n = \dim(\mathbf{F}_q^n)$,

$$k = n - (\ell(\mathbf{D}) - \ell(\mathbf{D} - \mathbf{B})),$$

and in particular if $2g - 2 < d(\mathbf{D})$, *then*

$$k = n - d(\mathbf{D}) + g - 1 + \ell(\mathbf{D} - \mathbf{B}),$$

and the minimal distance d *satisfies*

$$d \geq d(\mathbf{D}) - 2g + 2.$$

Proof. It follows from the previous discussion that

$$k = i(\mathbf{D} - \mathbf{B}) - i(\mathbf{D}).$$

It follows from the Riemann–Roch theorem that

$$i(\mathbf{D}) = \ell(\mathbf{D}) - d(\mathbf{D}) - 1 + g,$$

$$i(\mathbf{D} - \mathbf{B}) = \ell(\mathbf{D} - \mathbf{B}) - d(\mathbf{D} - \mathbf{B}) - 1 + g.$$

Therefore, we have

$$k = -\ell(\mathbf{D}) + d(\mathbf{D}) + \ell(\mathbf{D} - \mathbf{B}) - d(\mathbf{D} - \mathbf{B}))$$

$$= n - (\ell(\mathbf{D}) - \ell(\mathbf{D} - \mathbf{B})).$$

In case that $2g - 2 < d(\mathbf{D})$, we use Proposition 4.44 Riemann–Roch theorem to conclude that $i(\mathbf{D}) = 0$ and $\ell(\mathbf{D}) = d(\mathbf{D}) + 1 - g$. Therefore, we have

$$k = n - (\ell(\mathbf{D}) - \ell(\mathbf{D} - \mathbf{B}))$$

$$= n - d(\mathbf{D}) + g - 1 + \ell(\mathbf{D} - \mathbf{B}).$$

Let us assume that there are more than $n - (d(\mathbf{D}) - 2g + 2)$ zero-residues for a differential η at $\{\mathbf{P}_1, \ldots, \mathbf{P}_n\}$, that is, under the residue map, $\mathrm{res}(\eta)$ has more than $n - (d(\mathbf{D}) - 2g + 2)$ points of $\{\mathbf{P}_1, \ldots, \mathbf{P}_n\}$ with zero values. It thus has less than $(d(\mathbf{D}) - 2g + 2)$ points of $\{\mathbf{P}_1, \ldots, \mathbf{P}_n\}$ with non-zero values. We wish to prove that $\eta = 0$.

Let $\mathbf{B}' = \sum_j \mathbf{P}_j$, where j runs through all points $\mathbf{P}_j$ with $\mathrm{res}_{p_j} \eta = 0$. Note that η will have no poles at those $\mathbf{P}_j$. Then, as an assumption, we have $d(\mathbf{B}') > n - (d(\mathbf{D}) - 2g + 2)$ or $d(\mathbf{B} - \mathbf{B}') < d(\mathbf{D}) - 2g + 2$. Let us write

$\mathbf{D} = \mathbf{D_0} - \mathbf{D_\infty}$ as both effective and disjoint. We have $d(\mathbf{D}) = d(\mathbf{D_0}) - d(\mathbf{D_\infty}) = r - s$ (for notation, see the proof of Proposition 5.2). Then, we have $(\eta)_0 - (\eta)_\infty \succ \mathbf{D_0} - \mathbf{D_\infty} - \mathbf{B}$, $d((\eta)_0) \geq d(\mathbf{D_0}) = r, \mathbf{D_\infty} + (\mathbf{B} - \mathbf{B'}) \succ (\eta)_\infty$, and $d((\eta)_\infty) < s + (d(\mathbf{D}) - 2g + 2) = r - 2g + 2$. Therefore, we have $d((\eta)) = d((\eta)_0) - d((\eta)_\infty) > 2g - 2$. If $\eta \neq 0$, then $d((\eta)) = 2g - 2$. Hence, $\eta = 0$. What we have proved is that any non-zero differential η cannot have more than $n - (d(\mathbf{D}) - 2g + 2)$ zero-residues. Therefore, it has more than or equal to $(d(\mathbf{D}) - 2g + 2)$ non-zero-residues. It means that we prove

$$d \geq d(\mathbf{D}) - 2g + 2. \qquad \Box$$

Remark 3: The lower bounds for the rank $(n - d(\mathbf{D}) + g - 1 + \ell(\mathbf{D} - \mathbf{B}))$ and the minimal distance $(d(\mathbf{D}) - 2g + 2))$ are called the **designed rank** and **designed minimal distance**. ∎

Example 3: Recall Definition 3.20 of classical Goppa code as follows.

Definition 3.20. Let $\mathbf{G} = \{\gamma_1, \ldots, \gamma_n\} \subset \mathbf{F}_q$. We define the (*classical*) Goppa code $\Gamma(\mathbf{G}, g(x))$ with *Goppa polynomial* $g(x)$, where $g(\gamma_i) \neq 0$ for $1 \leq i \leq n$ to be the set of code words $\mathbf{c} = [c_1, \ldots, c_n]$ over the letter field $\mathbf{F}_q$ for which

$$\sum_{i=1}^{n} \frac{c_i}{(x - \gamma_i)} \equiv 0 \mod (g(x)).$$

∎

Let us consider a *geometric Goppa code* as follows. Let the curve C be the projective line $P^1_{F_q}$ over a finite field $\mathbf{F}_q$. Let $\beta_1, \ldots, \beta_n$ be n distinct elements in F_q (Note that $n \leq q$). Let $g(x)$ be a polynomial with $g(\beta_j) \neq 0$ for all j. Let us consider $\mathbf{P}_j$ equals the point defined by $x - \beta_j = 0$ and $\mathbf{P}_\infty$ equals the point at ∞. Let

$$\mathbf{D} = (g(x))_0 - \mathbf{P}_\infty = \sum_{\mathbf{Q} \neq \mathbf{P}_\infty} \mathrm{ord}_{\mathbf{Q}}(g(x))\mathbf{Q} - \mathbf{P}_\infty$$

and $\mathbf{B} = \sum \mathbf{P}_i$. Let us pick any non-zero $\eta \in \Omega(\mathbf{D} - \mathbf{B})$. Then,

$$\eta = f(x)dx = \frac{r(x)}{s(x)}dx.$$

Since η has only poles at finite distance in $\mathbf{B}$, then $s(x) \mid \prod(x - \beta_j)$. We may assume that $s(x) = \prod(x - \beta_i)$. Check the order at $\mathbf{P}_\infty$. Note that given $\deg(r(x)) = m, \deg(s(x)) = n$, the number of poles at ∞ for $r(x)/s(x)$ is $m - n$, and dx has two poles at ∞. Since η has at most one pole at ∞, we have $m - n + 2 \leq 1$ or $m \leq n - 1$ or $\deg(r(x)) \leq n - 1$. So, we have by the theory of partial fractions that

$$\eta = f(x)dx = \sum \frac{c_j}{x - \beta_j} dx.$$

Then, clearly c_j = residue at $\mathbf{P}_j$ and the mapping which maps η to $[c_1, \ldots, c_n]$ are the same for the residue map $\eta \;:\mapsto\; [\mathrm{res}_{P_1}, \ldots, \mathrm{res}_{P_n}]$. Furthermore, $\mathrm{ord}_\mathbf{Q}(r(x)) \geq \mathrm{ord}_\mathbf{Q}(g(x))$ for all $\mathbf{Q}$. Therefore, $g(x) \mid r(x)$. We conclude that $\eta \in \Omega(\mathbf{D} - \mathbf{B}) \Leftrightarrow \sum \frac{c_i}{x - \beta_i} \equiv 0 \mod (g(x))$. It is easy to see that this is the **classical Goppa Code**. ∎

Both Reed–Solomon codes and classical Goppa codes lead to geometric Goppa codes. We have a third definition of geometric Goppa codes as follows.

Definition 5.5 (Algebraic geometric code or a geometric Goppa code in primary form $C_p(\mathbf{B}, \mathbf{D})$). Let C be an absolutely irreducible smooth projective algebraic curve of genus g over $\mathbf{F}_q$. Consider an (ordered) set $\{\mathbf{P}_1, \mathbf{P}_2, \ldots, \mathbf{P}_n\}$ of distinct $\mathbf{F}_q$ rational points on C and a divisor $\mathbf{B} = \sum_{i=1}^n \mathbf{P}_i$ and an $\mathbf{F}_q$-divisor $\mathbf{D}$ on C. For simplicity, let us assume that the support of $\mathbf{D}$ is disjoint from $\mathbf{B}$. The *primary Goppa code* is defined as the set of vectors $[a_1, \ldots, a_n] \in \mathbf{F}_q^n$ such that $\sum a_j f(\mathbf{P}_j) = 0$ for all $f \in L(\mathbf{D})$. ∎

We wish to establish the relations between the three different versions of *geometric Goppa codes* by proving the following propositions.

Proposition 5.6. *The geometric Goppa code in function form $C_L(\mathbf{B}, \mathbf{D})$ is the dual code of the geometric Goppa code in primary form $C_p(\mathbf{B}, \mathbf{D})$. The dimension of the code space of the geometric Goppa code in primary form $C_p(\mathbf{B}, \mathbf{D})$ is $n - (\ell(\mathbf{D}) - \ell(\mathbf{D} - \mathbf{B}))$.*

Proof. It is evident from the definitions.   □

We consider two codes are identical if both their word spaces are the same $\mathbf{F}_q^n$ and their code spaces are the same subspace.

Proposition 5.7. *The* geometric Goppa code in residue form $C_\Omega(\mathbf{B}, \mathbf{D})$ *is the* geometric Goppa code in primary form $C_p(\mathbf{B}, \mathbf{D})$.

Proof. We prove the code space of $C_\Omega(\mathbf{B}, \mathbf{D}) \subset$ the code space of $C_p(\mathbf{B}, \mathbf{D})$. Then, by a dimension argument, since they are both of dimensions $n - (\ell(\mathbf{D}) - \ell(\mathbf{D} - \mathbf{B}))$, they must be equal.

Let $f \in L(\mathbf{D})$ and $\eta \in \Omega(\mathbf{D} - \mathbf{B})$. Then, $(f) \succ -\mathbf{D}$ and $(\eta) \succ \mathbf{D} - \mathbf{B}$. It follows that $(f\eta) = (f) + (\eta) \succ -\mathbf{B}$. It means that all residues of $f\eta$ outside $\{\mathbf{P}_j\}$ are zeroes. So, we have (cf. Proposition 4.37)

$$\sum f(\mathbf{P}_j) \operatorname{res}_{\mathbf{P}_j}(\eta) = 0.$$

Hence, we conclude that the code space of $C_\Omega(\mathbf{B}, \mathbf{D})$ and $C_p(\mathbf{B}, \mathbf{D})$ are each the dual code of the code space of $C_L(\mathbf{B}, \mathbf{D})$. Hence, the *geometric Goppa code in residue form* $C_\Omega(\mathbf{B}, \mathbf{D})$ is the *geometric Goppa code in primary form* $C_p(\mathbf{B}, \mathbf{D})$. $\qquad\square$

Proposition 5.8. *Let C be a smooth projective curve over $\mathbf{F}_q$ with distinct rational places $\mathbf{P}_1, \ldots, \mathbf{P}_n, \mathbf{P}$ and $\mathbf{B} = \sum_{j=1}^n \mathbf{P}_j$; then, there exists a differential ω such that $(\omega) \succ -B$ such that*

$$\operatorname{res}_{\mathbf{P}_j}(\omega) = 1, \quad \textit{for } j = 1, \ldots, n.$$

Let $\mathbf{U} = \mathbf{B} + (\omega)$ or $(\omega) = \mathbf{U} - \mathbf{B}$. Clearly, the supports of $\mathbf{U}$, $\mathbf{B}$ are disjoint. Let $\mathbf{D}$ be any divisor with the support of $\mathbf{D}$ disjoint from the support of $\mathbf{B}$. Let $\mathbf{D}' = \mathbf{U} - \mathbf{D}$ (on the other hand, if we are given any divisor $\mathbf{D}'$ with the support of $\mathbf{D}'$ disjoint from the support of $\mathbf{B}$, we may define $\mathbf{D} = \mathbf{U} - \mathbf{D}'$). Then, the geometric Goppa code in residue form $C_\Omega(\mathbf{B}, \mathbf{D})$ is isomorphic to the geometric Goppa code in function form $C_L(\mathbf{B}, \mathbf{D}')$.

Proof. Clearly, it follows from the **remark** after Proposition 4.38 that the differential ω exists. The divisor $(\omega) = -\mathbf{B} + \mathbf{U}$, such that the support of $\mathbf{B}$ is disjoint from the support of $\mathbf{U}$. Let us define a map $\pi : C_L(\mathbf{B}, \mathbf{D}') \mapsto C_\Omega(\mathbf{B}, \mathbf{D})$ by

$$\pi(f) = f\omega.$$

It is easy to see that

$$f \in L(\mathbf{D}') \Leftrightarrow (f) + \mathbf{U} - \mathbf{D} \succ 0$$
$$\Leftrightarrow (f) \succ \mathbf{D} - \mathbf{U} \Leftrightarrow (f) + (\omega) \succ \mathbf{D} - \mathbf{U} + \mathbf{U} - \mathbf{B} = \mathbf{D} - \mathbf{B}$$
$$\Leftrightarrow f\omega \in \Omega(\mathbf{D} - \mathbf{B}).$$

Furthermore, $\pi : L(\mathbf{D}' - \mathbf{B}) \mapsto \Omega(\mathbf{D})$ by

$$f \in L(\mathbf{D}' - \mathbf{B}) \Leftrightarrow (f) + \mathbf{U} - \mathbf{D} - \mathbf{B} \succ 0 \Leftrightarrow (f) \succ \mathbf{B} + \mathbf{D} - \mathbf{U}$$
$$\Leftrightarrow (f) + (\omega) \succ \mathbf{B} + \mathbf{D} - \mathbf{U} + \mathbf{U} - \mathbf{B} = \mathbf{D} \Leftrightarrow f\omega \in \Omega(\mathbf{D}).$$

Hence, $\pi : L(\mathbf{D}')/L(D' - B) \mapsto \Omega(\mathbf{D} - \mathbf{B})/\Omega(\mathbf{D})$, and π induces an isomorphism. $\qquad\square$

This theory is sensitive to the curves and divisors involved. For instance, there are curves with very few rational points or even no rational points. Clearly, the theory is bad or void in that case. For smooth plane curves, since the plane P_K^2 has $q^2 + q + 1$ rational points, there are at most $q^2 + q + 1$ rational points on the plane curve, and the *geometric Goppa code* based on it will be short. Or the divisor selected is poor for the application purposes (we have to take the decoding process in consideration). Therefore, we shall consider space curves. We are interested in special curves with many rational points so that the selection is easy with special divisors which may aid us in decoding.

We illustrate the above method by several examples as follows.

Example 4: Let us fix the ground field to be $\mathbf{F}_{2^4}$. Let us consider curve C defined by $x_0^5 + x_1^5 + x_2^5$ over $\mathbf{F}_{2^4}$. It follows from **Example 35** in Section 4.9 that the number of rational points is 65 and the genus $g = 6$. Let us take one rational point $\mathbf{P}$, $\mathbf{B} =$ the sum of the remaining 64 points and $\mathbf{D} = 37\mathbf{P}$. Then, $d(\mathbf{D} - \mathbf{B}) = -27 < 0$, and $\ell(\mathbf{D} - \mathbf{B}) = 0$. Note that $10 = 2g - 2 < d(\mathbf{D}) = 37$. Therefore, for *geometric Goppa code* in *residue form* (cf. Proposition 5.4), the rank $k = 64 - 37 + 6 - 1 = 32$, and the minimal distance $d \geq 27$. As for *Goppa code* in *function form*, the rank $k = 32$, and the minimal distance $d \geq 27$. We have the following data: $n = 64$, $k = 32$, the rate of information is $k/n = 32/64 = 1/2$, and the number of maximal correctable errors is $(d-1)/2 \geq 13$. Comparing with the **Reed–Solomon** code over F_{2^4}, which has $n = 2^4 - 1 = 15$, $1 \leq k < 15$, say $k = 7$, then the rate of information is closed to $1/2$, and $d = n - k + 1 = 9$. It may correct 4 errors. It is a much shorter code with weaker correcting power. ∎

5.2. Comparisons with Reed–Solomon Code

As pointed out by the **remark** of Proposition 1.25, any finite code, linear or not, can be decoded by brute force. In the next two chapters, we discuss two faster ways of decoding *algebraic curve Goppa codes* and compare them with a brute-force decoding. In this section, we show that if two codes are with closed rates of information and closed rates of distances, then the longer one is more precise. This is easy to understand. For instance, let C_1 be an $[n, k, d]$ code which corrects $\lfloor \frac{d-1}{2} \rfloor$ errors and C_2 be an $[2n, 2k, 2d]$ code which corrects $\lfloor \frac{2d-1}{2} \rfloor$. Let them work on a block of length $2n$. Then, we use a decoder for C_1 twice and a decoder for C_2 once. Then, any received word which can be decoded by the decoder for C_1 can be done by the decoder for C_2. On the other hand, a received word with $\lfloor \frac{d-1}{2} \rfloor + 1$ errors for the first n letters can only be decoded by the second decoder if the total number of errors is less than or equal to $\lfloor \frac{2d-1}{2} \rfloor$. Hence, it is easy to see that C_2 is more precise. We show this point by simple computations.

The decoding process for Reed–Solomon codes are faster than the known decoding algebraic geometry Goppa codes (see the next chapter). However, the algebraic geometry Goppa codes are more precise than the Reed–Solomon codes. The ground field controls the time required for multiplication, which certainly affects the total speed of computation. For our comparisons of different codes we shall fix a ground field. Let us fix the ground field to be $\mathbf{F}_q = \mathbf{F}_{p^m}$; there are $q + 1$ smooth points in $P^2_{\mathbf{F}_q}$. There, the length of code is $\leq q - 1 = 15 = n$ for Reed–Solomon codes, classical Goppa codes, etc. The Weil's theorem on the number of smooth rational points of a projective curve gives us a much bigger number, and Example 4 of the preceding section shows that a projective curve C over $\mathbf{F}_{2^4}$ could have 64 points, which is many more that $2^4 + 1 = 17$ points. So, we have a longer code. We discuss why a longer code is more precise.

We follow après Pretzel [8], p. 69. A further advance of *geometric Goppa codes* in the future might be an improvement of the speed of decoding. If the speed is not an issue, then the *geometric Goppa code* has an advantage of being more accurate, which will be illustrated in this section.

Let us consider a Reed–Solomon code over the field F_{2^4} defined by all polynomials of degree < 7. Then, it is a $[15, 7, 9]$-code, where $15 = 2^4 - 1, 7 = 7, 9 = 15 - 7 + 1$. So, the number of message symbols is 7, and it can correct $4 = (9 - 1)/2$ errors. Let us consider a *geometric Goppa code* based on the curve $x_0^5 + x_1^5 + x_2^5$ (cf. **Example 4** in the preceding section). By Plücker's formula, its genus g is 6. We have computed to find that there are 65 rational points (cf. **Example 35** in Section 4.9). Let one of these 65 points be **P**. Let us consider a one-point code with **B** = the sum of other 64 points and **D** $= 37\mathbf{P}$. Then, it is an $[n, k, d]$-code, where $n = 64, k = 32, d \geq 27$. Using *SV* algorithm (see the following), it may correct $(27 - 6 - 1)/2 = 10$ errors. Let us process through the said Reed–Solomon-code four times. Then, we process 28 message symbols, while we may process 32 symbols through the said *geometric Goppa code*.

Let us compare these two processors. We have $32 > 28$ and *geometric Goppa code* carrying more messages. Another yardstick is the failure rate, i.e., the probability that one may fail to recover code messages due to there being more than allowed errors and mistakenly decoding to wrong messages. To simplify our notations, we say that the decoder always returns an *error message* in those cases. Let us compute the probability for each processors to return an *error message*, i.e., if there are more than allowed errors appearing. For Reed–Solomon process, if there are 5 or more errors, then the processor will not decode and will reject the received word. Let the channel has a probability p of being incorrect and q of being correct. Then, $p + q = 1$ and

$$1 = 1^n = (p+q)^n = \sum_{i=0}^{n} \mathbf{C}_i^n p^i q^{n-i}$$

for a plain processor without self-correcting capability, and the probability of returning an *error message* is

$$r_0 = \sum_{i=1}^{1} p^i = p = 1 - q.$$

For Reed–Solomon processor, the probability r_1 of returning an *error message* is

$$r_1 = \sum_{i=5}^{15} \mathbf{C}_i^{15} p^i q^{15-i} = 1 - \sum_{i=0}^{4} \mathbf{C}_i^{15} p^i q^{15-i}.$$

By the same arguments, using *SV* algorithm for *geometric Goppa code*, the probability r_2 of returning an *error message* is

$$r_2 = \sum_{i=11}^{64} \mathbf{C}_i^{64} p^i q^{64-i} = 1 - \sum_{i=0}^{10} \mathbf{C}_i^{64} p^i q^{64-i}.$$

Let us assign a numeral for p. Let $r_0 = p = 0.01$. Then, the above numbers are $r_0 = 0.01$ and $r_1 = 0.27627423 \times 10^{-6}$, and $r_2 = 0.595098292 \times 10^{-10}$. Therefore, to use the Reed–Solomon processor four times, the probability R_1 of returning an *error message* is

$$R_1 = 1 - (1 - r_1)^4 = 0.11052 \times 10^{-5}.$$

We have $R_2 = r_2 = 0.595098292 \times 10^{-10}$. Therefore, the *SV algorithm for geometric Goppa code* is more accurate than the Reed–Solomon-code. We may consider a block of 70,000 message symbols. We have to use the plain processor 70,000 times, the Reed–Solomon processor 10,000 times, and the *geometric Goppa code* 2,188 times. The probabilities of *error message* for these three processors are

$$1 - (1 - r_0)^{70000} = 100\% \; for \; plain \; processor,$$

$$1 - (1 - r_1)^{10000} = 0.27591868\% \; for \; Reed - -Solomon, and$$

$$1 - (1 - r_2)^{2188} = 0.00002188\% \; for \; SV \; algorithm \; for \; Goppa \; code.$$

Later, in Section 6.4, we discuss the *DU* algorithm. Using it, we may correct up to 13 errors for this particular *geometric Goppa code*. Then, we have

$$r_3 = \sum_{i=14}^{64} \mathbf{C}_i^{64} p^i q^{64-i} = 1 - \sum_{i=0}^{13} \mathbf{C}_i^{64} p^i q^{64-i} = 0.1383190860 \times 10^{-10}$$

and $1 - (1 - r_3)^{2188} = 0.000000000655500732\%$, which is much better than the Reed–Solomon code.

5.3. Improvement of Gilbert–Varshamov's Bound

It follows from Proposition 1.33 (**Shannon's theorem**) that we may have to consider a sequence of codes with lengths tending to ∞ to achieve an asymptotically good result. On the other hand, Proposition 1.40 (Gilbert–Varshamov's bound) establishes a lower bound,

$$\alpha(\delta) \geq 1 - H_q(\delta).$$

In this section, we show that the above Gilbert–Varshamov's bound can be improved by a family of *geometric Goppa codes*. We consider *geometric Goppa codes* based on a sequence of curves C_ℓ with $\ell \mapsto \infty$. We require that the numbers n_ℓ of rational smooth points of C_ℓ tends to ∞ to satisfy the requirement of Proposition 1.33 (Shannon's theorem), Then, it follows from Proposition 4.49 that the genus g_ℓ must tend to ∞. Certainly, this consideration is only for theoretical purposes. For any application, we can only allow ℓ to become as large as possible.

5.3.1. *Modular Curves and Shimura Modular Curves*

For the theoretical part of coding theory, we are interested in a sequence of curves. There are well-known sequences of algebraic curves called as the *modular curves* which are discussed briefly in this section.

From *complex analysis*, we view an *elliptic curve* $\mathcal{E}$ as a torus, which can be represented as the quotient space of C/L, where the lattice L can be defined as a *rank 2* $\mathbb{Z}$*-submodule* of the complex numbers $\mathbb{C}$ generated by two complex numbers ω_1, ω_2 such that ω_1/ω_2 is not a real number. We may choose the order of ω_1, ω_2 such that $im(\omega_1/\omega_2) > 0$. Then, ω_1', ω_2' define the same lattice L iff

$$\begin{pmatrix} a, & b \\ c, & d \end{pmatrix} \begin{pmatrix} \omega_1 \\ \omega_2 \end{pmatrix} = \begin{pmatrix} \omega_1' \\ \omega_2' \end{pmatrix},$$

where the two-by-two matrix above is in the *full modular group* = the special linear group over $\mathbb{Z} = SL_2(\mathbb{Z}) = \Gamma$. Let the *principal congruence subgroup* of level N be

$$\Gamma(N) = \left\{ \begin{pmatrix} a, & b \\ c, & d \end{pmatrix} : det \begin{pmatrix} a, & b \\ c, & d \end{pmatrix} = 1, \ a \equiv d \equiv 1, \ b \equiv c \equiv 0 \mod N \right\}.$$

Two lattices L, L' induce isomorphic elliptic curves iff $L = c \cdot L'$, where $c \in \mathbb{C}^* = \mathbb{C}\backslash\{0\}$. Then, we may normalize the pair (ω_1, ω_2) by multiplying ω_2^{-1} to the pair to get $(z, 1)$. Note that we require $im(z) > 0$. We have the following computation:

$$z' = \frac{\omega_1'}{\omega_2'} = \frac{a\omega_1 + b\omega_2}{c\omega_1 + d\omega_2} = \frac{az + b}{cz + d}.$$

We shall *define* (it to be unorthodox from the point of view of algebra:)

$$\alpha(z) = \begin{pmatrix} a, & b \\ c, & d \end{pmatrix} z = \frac{az + b}{cz + d}.$$

Let $\mathcal{H} = \{z \in \mathbb{C} : im(z) > 0\}$ be the upper half-plane of the complex plane. It is easy to see that (recall $ad - bc = 1$) $im(\alpha(z)) > 0$ as in the following computation:

$$im(z') = im\left(\frac{az + b}{cz + d}\right) = im\left(\frac{(az + b)(c\bar{z} + d)}{|cz + d|^2}\right)$$

$$= im\left(\frac{adz + bc\bar{z}}{|cz + d|^2}\right) = \frac{im(z)}{|cz + d|^2}.$$

Therefore, α sends elements in $\mathcal{H}$ to elements in $\mathcal{H}$.

Let $Y(N) = \Gamma(N)\backslash\mathcal{H}$ and endow it with the quotient topology. It is known that $Y(N)$ is an affine curve. It is possible to deduce that $Y(1)$ is the *moduli curve* of all isomorphic classes of elliptic curves. Let $X(N)$ be the compactification of $Y(N)$. Then, it is well known that $X(N)$ is a projective algebraic curve, and the sequence of $X(N)$ is a sequence of curves.

Very similar to the modular curves, we have the Shimura curves or Shimura modular curves. We shall only mention them in the simplest way. What we use in coding theory is the following Proposition 5.9.

Given a finite field F_{p^2}, let $\ell(\neq p)$ be a prime. Let $\Gamma_0(\ell)$ denote the following multiplicative subgroup of $SL_2(\mathbb{Z})$:

$$\Gamma_0(\ell) = \left\{ \begin{pmatrix} a, & b \\ c, & d \end{pmatrix} : det \begin{pmatrix} a, & b \\ c, & d \end{pmatrix} = 1, \text{ and } c \equiv 0 \mod \ell \right\}.$$

Then, $\Gamma_0(\ell)$ acts on the half-plane $\mathcal{H}$, and $\Gamma_0(\ell) \backslash \mathcal{H}$ is an affine curve $Y_0(\ell)$ (for a detailed discussion, see Tsfasman *et al.* [33]). We define the reduction of $Y_0(\ell)$ by characteristic p and complete it to a projective curve $X_0(\ell)(= C_\ell)$. Then, its genus is $\lfloor \ell/12 \rfloor$. We have the following proposition.

Proposition 5.9 (Ihara–Tsfasman–Vlăduţ–Zink). *For a finite field F_{p^2} with p prime, there exists a sequence of curves $C_\ell(= \text{Shimura modular curves } X_0(\ell))$ which have genus $g_\ell = \lfloor \ell/12 \rfloor$ and the number of rational points $n_\ell \geq (p-1)(\ell+1)/12$. Thus, $g_\ell/n_\ell \leq 1/(p-1)$.*

Proof. Ihara [24], Tsfasman–Vlăduţ–Zink [33]. □

Remark: The curves mentioned above are interesting. For instance, there is no smooth-plane models for them if ℓ is large enough. Note that a projective plane $P_{F_q}^2$ can be decomposed as $\mathbf{A}_{\mathbf{F_q}}^2 \cup \mathbf{A}_{\mathbf{F_q}} \cup 0$; its number of rational points is $q^2 + q + 1$. Therefore, a smooth plane curve has at most $q^2 + q + 1$ rational points; then, the Shimura curves $X_0(\ell)$ do not have a smooth and regular planar model if $\ell > (12(p^4 + p^2 + 1)/(p-1) - 1)$, where $q = p^2$. Note that then, we have the number of rational points $n_\ell \geq (p-1)(\ell+1)/12 > (p-1)12(p^4 + p^2 + 1)/((p-1)12) = (q^2 + q + 1)$ the number of rational points of a plane. ∎

5.3.2. *The Theorem of Tsfasman, Vlăduţ, and Zink*

We want to show that geometric Goppa codes can better than the well-known Gilbert–Varshamov's bound. Recall the following *entropy* function $H_q(x)$ with $H_q(0) = 0$ and defined for $0 < x < (q-1)/q$:

$$H_q(x) = x \log_q(q-1) - x \log_q(x) - (1-x) \log_q(1-x).$$

We have the following proposition which will be used in the next theorem.

Proposition 5.10. *The maximum value of $H_q(x) - x$ is attained for $x = (q-1)/(2q-1)(= \delta)$. The maximal value of $H_q(x) - x$ is $\log_q(2q-1) - 1$, i.e., we have $1 - H_q(\delta) = 2 - \delta - \log_q(2q-1)$.*

Proof. Using calculus, we differentiate $H_q(x) - x$ and set it to 0. We have

$$\log_q(q-1) - \log_q(x) - \log_q e + \log_q(1-x) + \log_q e - 1 = 0.$$

The above equation can be rewritten as

$$\log_q(q-1)(1-x)/(qx) = 0$$

or

$$(q-1)(1-x)/(qx) = 1.$$

Solving the above equation, we get

$$x = (q-1)/(2q-1)$$

and substituting it in $H_q(x) - x$, we have

$$(q-1)/(2q-1)\log_q(q-1) - (q-1)/2q - 1)\log_q(q-1)/(2q-1)$$
$$- (1 - (q-1)/(2q-1))\log_q(1 - (q-1)/(2q-1)) - (q-1)/(2q-1)$$
$$= \log_q(2q-1) - 1.$$

Using the second derivative test, we get

$$-\frac{\log_q e}{x} - \frac{\log_q e}{1-x} < 0 \text{ if } 0 < x < (q-1)/q.$$

We find it is negative, so we have the maximal point. □

We have the following theorem for coding theory.

Theorem 5.11. *Let us use the notations of the previous Proposition 5.9. Moreover, we assume that $p \geq 7$ and $q = p^2$. Then, we have a one-point code on $X_0(\ell)$ with block length $n_\ell - 1$ tending to ∞ such that for $\delta = (q-1)/(2q-1)$, the relative minimum distance $d_\ell/(n_\ell - 1)$ tends to a limit $> \delta$, and their rate of information $k_\ell/(n_\ell - 1)$ tends to a limit $> 1 - H_q(\delta)$.*

Proof. Let us consider the sequence of Shimura curves $X_0(\ell)$. Let us select one rational point **P** and call the other rational points $\mathbf{P}_1, \ldots, \mathbf{P}_{n_\ell-1}$.

Let $\mathbf{B} = \sum_{i=1}^{n_\ell - 1} \mathbf{P}_i$. Let $\mathbf{D} = \alpha_\ell \mathbf{P}$ with $\alpha_\ell > 2g_\ell - 2$. Then, the rank k_ℓ and minimal distance d_ℓ satisfies

$$k_\ell \geq (n_\ell - 1) - \alpha_\ell + g_\ell$$

$$d_\ell \geq \alpha_\ell - 2g_\ell + 2$$

and the rate of distance δ_ℓ and the rate of information R_ℓ are given as

$$R_\ell = k_\ell/(n_\ell - 1) \geq (n_\ell - \alpha_\ell + g_\ell - 1)/(n_\ell - 1),$$

$$\delta_\ell = d_\ell/(n_\ell - 1) \geq (\alpha_\ell - 2g_\ell + 2)/(n_\ell - 1).$$

Choose α_ℓ so that $(\alpha_\ell - 2g_\ell + 2)/(n_\ell - 1) \mapsto \delta$. Then, the limit of δ_ℓ is at least δ, and the limit of rate of information R_ℓ (cf. Proposition 5.9) is at least

$$1 - \delta - \lim(g_\ell/n_\ell) \geq 1 - \delta - 1/(p - 1).$$

Under our assumption, $p \geq 7$, and we want to prove that $\log_{p^2}(2p^2 - 1) > p/(p-1)$ for all $p \geq 7$. For $p = 7$, we have $\log_{72}(2 \cdot 49 - 1) = 1.1755 > 7/6 = 7/(7 - 1)$, or the inequality $\log_{p^2}(2p^2 - 1) - p/(p - 1) > 0$ is satisfied. It is not hard to see that it is a monotonic increasing function. Therefore, we always have $\log_{p^2}(2p^2 - 1) > p/(p - 1) = 1 + 1/(p - 1)$ for all $p \geq 7$. Note that we use $q = p^2$, the above result can be rewritten as $\log_q(2q - 1) > p/(p - 1) = 1 + 1/(p - 1)$. It follows from the preceding Proposition 5.10 that (cf. Definition 1.36, Proposition 1.40)

$$\alpha(\delta) = \lim_{\ell \to \infty} R_\ell \geq 1 - \delta - \lim(g_\ell/n_\ell) \geq 1 - \delta - 1/(p - 1)$$

$$> 2 - \delta - \log_q(2q - 1) = 1 - H_q(\delta). \qquad \square$$

Exercises

(1) Find the generator matrix for the code defined in **Example 2**.

(2) Find the check matrix for the code defined in **Example 2**.

(3) Write a computer program to code any message for the code defined in **Example 2**.

(4) Show that the Gilbert–Varshamov's bound is not reached by a sequence of Hamming codes.

Chapter 6

Decoding the Geometric Goppa Codes

6.1. Introduction

For a useful code, the decoding process is significant. **What do we mean by** *decoding* **is that there is an integer** t **such that given a received word** r**, (1) if there are less than or equal to** t **errors, then the decoder will find the original code word; (2) if there are more than** t **errors, then either we find a code word** c **which is within a distance** t **from the received word** r **(in general, if we find a code word, then with a small probability,** c **may not be the original sent code word) or we return an** *"error"* **message to indicate that what was found is not even a code word, and hence there are more than** t **errors.**

As we pointed out in the previous section (cf. Proposition 5.8), the three different forms of *geometric Goppa codes* determine each other. In the present chapter, we discuss only the decoding procedures for the *primary form*. Then, the decoding processes for the other two forms naturally follow.

Let us consider a primary code $C_p(\mathbf{B}, \mathbf{D})$ based on a smooth curve C over $\mathbf{F}_q$ of genus g, where $\mathbf{B} = \mathbf{P}_1 + \mathbf{P}_2 + \cdots + \mathbf{P}_n$ is a sum of distinct rational points and $\mathbf{D}$ is a divisor with the degree $d(\mathbf{D}) < n$ (cf. Proposition 5.2) and the support of $\mathbf{D}$ disjoint from $\mathbf{B}$.

A long message is chopped into many blocks of length n each. We shall only concentrate on decoding of one block. Hence, when we talk about messages, error words, words, etc., we always mean blocks of lengths n. Recall that the primary Goppa code is defined as the set of vectors $[a_1, \ldots, a_n] \in \mathbf{F}_q^n$ such that $\sum a_i f(\mathbf{P}_i) = 0$ for all $f \in L(\mathbf{D})$ whose dimension $k = n - \ell(\mathbf{D}) > 0$.

Let the minimal distance d of the code be $min\{d(a,b) : a \neq b$ are code words\}. Note that the only check we have to carry out for a vector $a = [a_1, \ldots, a_n]$ to be a code word is to check if $\sum_{i=1}^{n} a_i \phi_j(\mathbf{P}_i) = 0$ for a basis $\{\phi_j\}_{j=1}^{j=n-k}$ of the vector space $L(\mathbf{D})$. Later on, when we mention **check procedure**, we either mean to use the preceding procedure of check or use a **check matrix** $(\phi_j(\mathbf{P}_i))$ to check if $(\phi_j(\mathbf{P}_i))a^T = 0$, which are the same thing. Let d be the minimal distance of the code space.

The theoretical constructions of codes will be of interest to the general public if there are economical ways of decoding them. Although the **remark** after Proposition 1.25 points out that any code can be decoded by brute force if the code is finite, there are faster ways than that. We discuss the possible ways.

Most ways of decoding simply use matrix theory and hence are easy. There are two prominent ways, among many other methods, of decoding, which are the Skorobogatov–Vlăduţ algorithm (SV algorithm) (1991) and the Duursma algorithm (DU algorithm) (1993) which adopts the scheme of Feng and Rao of *majority voting* (1993).

The shortcomings of the above-mentioned two methods are the slow processes of solving linear equations. In general, they require n^3 (if we factor in some extra burden) steps to solve the system of linear equations which are comparably slower than the process of decoding Reed–Solomon code (n^2 steps). On the other hand, we already know that the *geometric Goppa codes* are more accurate than the *Reed–Solomon* codes (cf. Section 5.2) which is the advantage of *geometric Goppa codes*.

In the decoding procedures, let a non-negative number t be the number of errors to be corrected. Later, in the SV algorithm, t is $\leq \lfloor \frac{d-g-1}{2} \rfloor$, and in the DU algorithm, t is $\leq \lfloor \frac{d-1}{2} \rfloor$. The decoding procedures of correcting t errors (cf. the **remark** after Proposition 1.25) consist, in general, of three steps as in the following *Flowchart 2*. In *Flowchart 2*, **syndrm cal** means *syndrome calculation* (see the section on syndrome calculation under Section 6.2), **error e** means *error vector e*, and **error** means *error message*, which means that there are more than t errors and hence the decoder fails.

If the word r already goes through the **check** procedure (left half of the flowchart) and fails, then let it go through the **syndrm cal**. If it passes the **syndrm cal**, which means it either has no error or more than t errors, since it already failed the **check**, and it must have more than t errors, then it goes directly to **error**.

(1) The sender and the receiver agree on the *generator matrix* and the *check matrix* (cf. Exercises 4 and 5). Upon receiving the received word,

the receiver either (A) lets it goes to the left to check if it is a code word by using the *check matrix* or the **check** procedure to find if the received message is a code word. If it is a code word (it may not be the original sent code word), then go to the next block of message. If not, then use the following *syndrome calculation*, which lead it to the *syndrome calculation* (see Section 6.2) of the right column. Or (B) We may go to the right column directly. We apply the syndrome calculation to the received word r. There are two possibilities: either it passes the calculation or it fails the calculation. If it passes the calculation and comes from the left column with a failure for the check procedure or check matrix, that means the basic assumptions $1 \leq wt(e) \leq t$ is not true; therefore, we conclude that there are more than t errors, we return an *error* message. Or maybe, it directly comes to this calculation and passes, we have to check it by the checking procedure. If it further passes the check, then it is a code word; if it fails then it goes to *SV or DU*.

(2) We start either the SV algorithm (Section 6.3) or DU algorithm (Section 6.4), which is with the basic *assumption* that there are t or less errors. At the end of the procedures, we construct an error vector e.

(3) Further, the test will decide if $r - e$ is a code vector. If it is, then we complete the decoding procedure and correct the errors (occasionally, it may correct more than t errors) successively. If it is not, then there are more than t errors and return a message "error".

We have the following Figure 6.1.

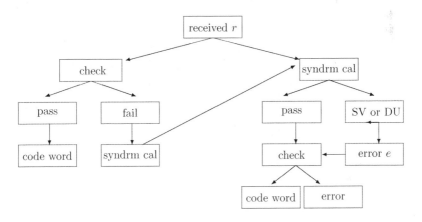

Figure 6.1. Decoding flowchart.

6.2. Error Locator

Let us consider a primary code $C_p(\mathbf{B}, \mathbf{D})$ based on a smooth curve C over $\mathbf{F}_q$ of genus g, where $q = 2^m$, $\mathbf{B} = \mathbf{P}_1 + \mathbf{P}_2 + \cdots + \mathbf{P}_n$ is a sum of distinct rational points and $\mathbf{D}$ is a divisor with the degree $d(\mathbf{D}) < n$ (cf. Proposition 5.2) and the support of $\mathbf{D}$ disjoint from $\mathbf{B}$. Let a code word c be transmitted and a word $r = c + e$ be received. The word e is called the *error* word. The decoding process is to find e with $1 \le wt(e) \le t$, given only r. Then, we find $c = r - e$.

Recall that the **check** procedure is to compute $\sum \phi_k(\mathbf{P}_j)r_j$ for $\{\phi_k\}$ a basis for $L(D)$. Given any function ϕ, those numbers $\phi(\mathbf{P}_j)r_j$ for $j = 1, \ldots, n$ are our main tools of searching. We define the following.

Definition 6.1. For any rational function $\phi, r \in$ the rational function field of C with no pole at $\mathbf{B}$, we define pseudo-dot (or a pairing) product $\cdot$ as

$$\phi \cdot r = \sum_j \phi(\mathbf{P}_j)r(\mathbf{P}_j) = \sum_j \phi(\mathbf{P}_j)r_j,$$

where $r_j = r(\mathbf{P}_j)$ and call $\phi \cdot r$ the *syndrome* of r with respect to ϕ. Note that it is a number $\in \mathbf{F}_q$. Let $\mathbf{U}$ be a divisor. We say that r has no *syndrome* with respect to $L(\mathbf{U})$, if $\phi \cdot r = 0$ for all $\phi \in L(\mathbf{U})$. It means that if there is no syndrome with respect to $L(\mathbf{D})$, then r is a code word. We define the $\odot$ product as

$$\phi \odot r = [\phi(\mathbf{P}_1)r(\mathbf{P}_1), \ldots, \phi(\mathbf{P}_n)r(\mathbf{P}_n)] = [\phi(\mathbf{P}_1)r_1, \ldots, \phi(\mathbf{P}_n)r_n],$$

which is a word. ∎

Remark: It is easy to check that $(\psi \odot \chi) \odot e = (\chi \odot \psi) \odot e = \chi \odot (\psi \odot e)$. ∎

We first define an **error locator** theoretically, and then find the properties of the error locator, and use the properties to locate the error locations. Once we get the error locator, we use it to find all error locations and the values for the error vector e at the locations. In this way, we find the original code word $c = r - e$. It will be of great help if we know all locations of the error vector e. Even a partial knowledge of the error locations will be meaningful. We define the following:

Definition 6.2. Given a received word r with the unknown original code word c. Let $e = (e_1, \ldots, e_n)$ be an error word with $c = r + e$. The point $\mathbf{P}_j$ is called an **error location** (with respect to e) if $e_j \ne 0$. A non-zero function

θ is called an **error locator** (with respect to e) if $\theta(\mathbf{P}_j) = 0$ for all error locations $\mathbf{P}_j$ and θ has no poles among $\mathbf{P}_1, \ldots, \mathbf{P}_n$. Thus, $\theta \odot e = [0, \ldots, 0]$ always. ∎

From an error locator θ, we cannot determine *the set of all error locations* M of e since θ may be zero outside M. However, if the set $M' = \{\mathbf{P}_j : \theta(\mathbf{P}_j) = 0\}$ is not too big (see Proposition 6.6) comparing with M (we have $M' \supset M$), then we show later (see Proposition 6.8) that it is possible to determine M and all coordinates $\{e_j : e_j \neq 0\}$ and solve the decoding problem.

6.2.1. *Plan of Decoding*

Let us assume that the received word r fails the test of **syndrome calculations**; we further assume that $1 \leq wt(e) \leq t$. Then, we want to find an error locator. If we can find a non-zero solution $\{\alpha_i\}$ of a system of equations from the left as follows, then we determine an error locator $\theta = \sum \alpha_i \psi_i$ (see Proposition 6.7):

$$\begin{bmatrix} \alpha_0 \cdots\cdots \alpha_{v-1} \end{bmatrix} \cdot \begin{bmatrix} * & \cdots & \cdots & * \\ \cdots & S_{i,j} & \cdots & \cdots \\ \cdots & \cdots & \cdots & \cdots \\ * & \cdots & \cdots & * \end{bmatrix} = \begin{bmatrix} 0 \cdots\cdots 0 \end{bmatrix}.$$

It is possible that we can find only trivial solutions; it simply means that our basic assumption $1 \leq wt(e) \leq t$ is wrong, i.e., there is no error or there are more than t errors. We have to use the check procedure or the check matrix to decide if there is no error or if there are more than t errors. Even we find the error locator θ, the basic assumption $1 \leq wt(e) \leq t$ may still be wrong, and we may find a faulty error locator θ and hence a faulty error vector e. We have to use the check procedure or the check matrix to decide.

We always assume that $1 \leq wt(e) \leq t$. After we already find an error locator θ (which may be faulty), we want to find error vector e (which may be faulty). Let $M' = \{\mathbf{P}_k : \theta(\mathbf{P}_k i) = 0\}$. We solve a system of equations, where $\mathbf{P}_k \in M'$, from the right to determine an unique error vector e, where $*$'s are known numbers:

$$\begin{bmatrix} * & \cdots & \cdots & * \\ \cdots & & \cdots & \cdots \\ \cdots & \phi_j(\mathbf{P}_k) & \cdots & \cdots \\ * & \cdots & \cdots & * \end{bmatrix} \cdot \begin{bmatrix} E_0 \\ \cdot \\ \cdot \\ \cdot \\ E_{u-1} \end{bmatrix} = \begin{bmatrix} * \cdot r \\ \cdot \\ \cdot \\ * \cdot r \end{bmatrix}.$$

It is possible that the error vector e we find may not be the true one. It is due to that our basic assumption $1 \leq wt(e) \leq t$ is not satisfied. We have to use the check procedure or the check matrix to decide if e is the true one by checking if $r - e$ is a code word.

6.2.2. *The General Properties of Error Locators and Syndrome Calculations*

Let us consider $L(\mathbf{D})$. Let $\{\varphi_0, \ldots, \varphi_{\ell-1}\}$ be a basis of $L(\mathbf{D})$. Then, we may write $\{\varphi_i(P_j)r_j\}$ as a $\ell \times n$ matrix $(\varphi_i(P_j)r_j)$. Note that if every row adds up to 0, then $\varphi_i \cdot r = 0, \forall i$, so there is no *syndrome* with respect to $L(\mathbf{D})$ and $r = c$ or $e = 0$. The above is the *check procedure*. We use the following proposition to decide if a received word r has either more than t errors or no error (cf. the remark after Proposition 1.25). It is an important proposition to understand the decoding procedures; it shows that if we only allow t or less errors in any block, we shall only check all functions χ in $L(\mathbf{Y})$ with $d(\mathbf{Y}) \geq t + 2g - 1$ and the support($\mathbf{Y}$) disjoint from support($\mathbf{B}$) instead of check $L(\mathbf{D})$. Note that even if there is **no syndrome**, it may be due to the false assumption that the number of error is $\leq t$ (i.e., the true number of errors $> t$). We have to further test to see if it is a code word.

Proposition 6.3. *Consider a received word r with error word e satisfying $1 \leq wt(e) \leq t$. If for all functions χ in $L(\mathbf{Y})$ with $d(\mathbf{Y}) \geq t + 2g - 1$ and the support($\mathbf{Y}$) disjoint from support($\mathbf{B}$), $\chi \cdot e = 0$, then $e = 0$, i.e., there is no syndrome.*

Proof. Note that e is a code word in $C_p(\mathbf{B}, \mathbf{Y})$ which can be identified with the residue form $C_\Omega(\mathbf{B}, \mathbf{Y})$ (Proposition 5.7). Furthermore, it follows from Proposition 5.4 that the minimal weight of the non-zero code word e is $wt(e) \geq d(Y) - 2g + 2 \geq t + 2g - 1 - 2g + 2 = t + 1$. Then, either $e = 0$ or $d(e) \geq t + 1$, and hence, with the assumption that $wt(e) \leq t$, we have $e = 0$. □

6.2.3. *Assumption of t Errors and the Error Locators*

The following is a useful and important criterion for error locator θ. This proposition will be applied also in the **DU Algorithm** in the next section.

Proposition 6.4. *Assume that $1 \leq wt(e) \leq t$. Assume that $d(\mathbf{Y}) \geq t + 2g - 1$ and support($\mathbf{Y}$) is disjoint from the support($\mathbf{B}$). Let θ be a rational* *function without poles in* **B**. *Then,* θ *is an error locator for* $e \Leftrightarrow \theta\chi \cdot e = 0$ *for all* $\chi \in L(\mathbf{Y})$.

Proof. Let $e' = \theta \odot e = (\theta(\mathbf{P}_1)e_1, \ldots, \theta(\mathbf{P}_n)e_n)$. Note that the weight of $e' \leq t$ since e has at most t non-zero coordinates.

($\Rightarrow$) Since θ is an error locator, we have $e' = 0$. We have

$$\theta\chi \cdot e = \chi \cdot (\theta \odot e) = \chi \cdot e' = 0,$$

for all $\chi \in L(\mathbf{Y})$.

($\Leftarrow$) Since $\chi \cdot e' = \chi \cdot (\theta \odot e) = \theta\chi \cdot e = 0$ for all $\chi \in (\mathbf{Y})$, then e' is a code word in $C_p(\mathbf{B}, \mathbf{Y})$ which has a minimal weight $\geq d(\mathbf{Y}) - 2g + 2 \geq (2g + t - 1) - 2g + 2 = t + 1 > t$. Therefore, $e' = (0, 0, \ldots, 0)$, and θ is an error locator for e. $\qquad\square$

6.2.4. *Syndrome Calculations*

We have the following proposition for *syndrome calculations*.

Proposition 6.5. *Consider a received word* r *with error word* e *having* $wt(e) \leq t$. *If all functions* χ *in* $L(\mathbf{U})$ *with* $d(\mathbf{U}) \geq t + 2g - 1$ *and* $\mathbf{D} \succ \mathbf{U}$ *satisfy* $\chi \cdot r = 0$, *then* $e = 0$, *i.e., there is no syndrome and* $r \in$ *the code space.*

Proof. Since $\mathbf{D} \succ \mathbf{U}$, then $L(\mathbf{U}) \subset L(\mathbf{D})$, and $\chi \cdot c = 0$ for all $\chi \in L(\mathbf{D})$. Therefore, $\chi \cdot r = \chi \cdot e = 0$ for all $\chi \in L(\mathbf{U})$. We conclude that $e = 0$ by the Proposition 6.3. $\qquad\square$

It follows from the preceding proposition that if we only consider a smaller divisor $\mathbf{U}$ with $d(\mathbf{U}) \geq t + 2g - 1$ and $\mathbf{D} \succ \mathbf{U}$, then as long as the received word r satisfies the following **syndrome calculus**, where $\varphi_0, \ldots, \varphi_{w-1}$ form a basis for $L(\mathbf{U})$, we can only conclude that either r is a code word or there are more than t errors:

$$\begin{bmatrix} \varphi_0(\mathbf{P}_1) & \cdots & \cdots & \varphi_0(\mathbf{P}_n) \\ \cdots & \cdots & \cdots & \cdots \\ \cdots & \cdots & \cdots & \cdots \\ \varphi_{w-1}(\mathbf{P}_1) & \cdots & \cdots & \varphi_{w-1}(\mathbf{P}_n) \end{bmatrix} \cdot \begin{bmatrix} r_1 \\ \cdot \\ \cdot \\ r_n \end{bmatrix} = \begin{bmatrix} 0 \\ \cdot \\ \cdot \\ 0 \end{bmatrix}.$$

Indeed, we have the following. (1) If the equation is satisfied, i.e., the answer is the zero vector 0 on the right-hand side of the equation, then there is no *syndrome* with respect to $L(\mathbf{U}) \subset L(\mathbf{D})$. However, we can only conclude that either r is a code word or there are more than t errors.

If we used the **check** procedure or the check matrix before heading to decide that there are errors, then certainly, there are more than t errors. Otherwise, we complete a basis $\varphi_0, \ldots, \varphi_{w-1}$ of $L(\mathbf{U})$ to a basis $\varphi_0, \ldots, \varphi_{\ell-1}$, we have to use the remaining part of check procedure now to complete it. Thus, we can tell if r is a code word or there are more than t errors.

Let us go back to the *syndrome calculation* without **check** first. (2) If the equation is not satisfied, i.e., if the answer on the right-hand side is a non-zero vector, we *assume* that there are fewer than t errors and use the *syndrome table* to find an error locator and the error vector e (see Propositions 6.7 and 6.8). Certainly, we have to check if $r - e$ is a code word by using the check matrix or the **check** procedure. If $r - e$ is a code word, then we succeed in correcting the errors. If $r - e$ is not a code word, then there are more than t errors and the decoder fails. We need $d(\mathbf{U}) \times n$ (sometimes only $(2g + t - 1) \times n$ as in the case of one-point codes) multiplications for the *syndrome calculation* of every block.

Suppose an **error locator** θ is found. We wish to estimate the size of the set $M' = \{P_i : \theta(P_i) = 0\}$. The following proposition gives an estimation.

Proposition 6.6. *Let* $\mathbf{A}$ *be a divisor. Then, any function* $\theta \in L(\mathbf{A})$ *can have at most* $d(\mathbf{A})$ *points as zeroes outside the support of* $\mathbf{A}$.

Proof. Let us write the divisor:

$$(\theta) + \mathbf{A} = \sum_{\mathbf{Q}_i \in support(\mathbf{A})} n_i \mathbf{Q}_i$$

$$+ \sum_{\mathbf{Q}_j \notin support(\mathbf{A})} n_j \mathbf{Q}_j \equiv \mathbf{E} + \mathbf{F}.$$

If $\mathbf{F} \not\succ 0$ or $\mathbf{E} \not\succ 0$, then $\mathbf{E} + \mathbf{F} \not\succ 0$. It is impossible. Thus, $\mathbf{F} \succ 0$, $\mathbf{E} \succ 0$. We wish to show that $d(\mathbf{F}) \leq d(\mathbf{A})$. Suppose the contrary, $d(\mathbf{F}) > d(\mathbf{A})$. Then, $d(\mathbf{E} + \mathbf{F}) > d(\mathbf{A})$, and $d(\mathbf{A}) = d((\theta) + \mathbf{A}) > d(\mathbf{A})$. A contradiction. $\square$

Since we usually select divisor $\mathbf{A}$ with support $(\mathbf{A})$ disjoint from support $(\mathbf{B})$ which contains M', then we conclude that $d(\mathbf{A}) \geq ||M'||$. This is an important estimation of the size of M'.

6.3. SV Algorithm

This is the part of step (2) of Section 6.1 of this chapter. Let us consider a *geometric Goppa code in primary form* $\mathbf{C}_p(\mathbf{B}, \mathbf{D})$ based on a smooth projective curve C with genus g and a suitable divisors $\mathbf{B}$, $\mathbf{D}$. Let the rank $k = n - d(\mathbf{D}) + g - 1$ (note that according to Proposition 4.28 (Riemann's theorem), if $d(\mathbf{D}) \geq 2g - 1$, then we have $\ell(\mathbf{D}) = d(\mathbf{D}) + 1 - g$, and $k = n - \ell(\mathbf{D}) = n - d(\mathbf{D}) + g - 1$) and the minimal distance $d = d(\mathbf{D}) - 2g + 2$ and the number of permissible errors $t \leq \lfloor \frac{d-g-1}{2} \rfloor$. The SV algorithm requires many numerical inequalities. They are for (1) syndrome calculation to decide if the syndrome of r with respect to $\mathbf{U}$ is 0, (2) the existence of error locator θ, and (3) the usage of θ to compute the error vector e.

6.3.1. *Sufficient Relations*

We want to find auxiliary divisors $\mathbf{U}$, $\mathbf{A}$, $\mathbf{Y}$, $\mathbf{X}$ which satisfy the following **list of relations** between them: (1) $\mathbf{D} \succ \mathbf{U}$; (2) $\mathbf{D} \succ \mathbf{Y} + \mathbf{A} \succ \mathbf{X}$; (3) $d(\mathbf{U}) \geq t + 2g - 1$; (4) $\ell(\mathbf{A}) > t$; (5) $d(\mathbf{A}) < d(\mathbf{D}) - (2g - 2) - t$; (6) $d(\mathbf{X}) \geq d(\mathbf{A}) + 2g - 1$; (7) $d(\mathbf{Y}) \geq t + 2g - 1$; (8) support($\mathbf{Y}$) $\cap$ support($\mathbf{B}$) = support($\mathbf{A}$)$\cap$ support($\mathbf{B}$) = $\emptyset$.

By the preceding 6.5 proposition, $\mathbf{U}$, if exists, can be used for *syndrome* calculation. Do we know if those divisors $\mathbf{D}$, $\mathbf{U}$, $\mathbf{A}$, $\mathbf{Y}$, $\mathbf{X}$ exist? Let us consider a simple example of one-point code, we want to show that the above list of relations can be satisfied in this case. This shows that the complicated **sufficient relations** can be satisfied sometimes. After all, we often use an one-point code.

6.3.2. *One-Point Code*

Let us consider one-point codes of a special kind.

Let us consider a one-point code $\mathbf{C}_p(\mathbf{B}, m\mathbf{P})$, where $m \geq 3g + 2t - 1$. We take $\mathbf{D} = m\mathbf{P} = \mathbf{X}$, $\mathbf{A} = b\mathbf{P}$, with $g + t \leq b \leq m - 2g - t + 1$, $\mathbf{U} = \mathbf{Y} = (m - b)\mathbf{P}$.

We show that the above **sufficient relations** can be fulfilled. It is easy to check that for the above **list of relations** (1) $\mathbf{D} \succ \mathbf{U}$, (2) $\mathbf{D} \succ \mathbf{Y} + \mathbf{A} \succ \mathbf{X}$, and (8). Support($\mathbf{Y}$)$\cap \mathbf{B} = \emptyset$.

We use the following lemma to show that all numerical conditions are satisfied. For (3), $d(\mathbf{U}) = m - b = d(\mathbf{D}) - d(\mathbf{A}) \geq t + 2g - 1$. For (4), it follows from (i) of the following lemma. For (5), it follows from (ii) of the following lemma. For (6), it follows from (iii) of the following lemma since $\mathbf{X} = \mathbf{D}$. For (7), it follows from (iii) of the following lemma. For (8), it follows from our selections of $\mathbf{Y}$, $\mathbf{A}$, $\mathbf{B}$. It follows that all numerical conditions are satisfied.

Lemma. *If $0 \leq t$ (t is the number of errors we wish to correct) and $m = d(\mathbf{D}) \geq 3g + 2t - 1$, then there is a positive integer b such that*

$$t + g \leq b \leq d(\mathbf{D}) - 2g - t + 1. \tag{1}$$

Furthermore, since the divisor $\mathbf{A}$ is with $d(\mathbf{A}) = b$, then

(i): $\ell(\mathbf{A}) > t$,

(ii): $b = d(\mathbf{A}) < d(\mathbf{D}) - (2g - 2) - t = d(\mathbf{D}) - 2g - t + 2$,

(iii): $d(\mathbf{D}) - b \geq 2g + t - 1 > 2g - 2$.

Note that the *designed minimal distance* $d = d(\mathbf{D}) - 2g + 2$. Equation (1) means $t \leq \lfloor \frac{d-g-1}{2} \rfloor$.

Proof. Let us prove the last inequality first. Clearly, we have

$$2t \leq d(\mathbf{D}) - 3g + 1 \Leftrightarrow t + g \leq d(\mathbf{D}) - 2g - t + 1 = d - t - 1 \Leftrightarrow 2t \leq d - g - 1.$$

Then, it follows from (1) that

$$t \leq \left\lfloor \frac{d - g - 1}{2} \right\rfloor.$$

We just pick up an integer b in between $t + g$ and $d(\mathbf{D}) - 2g - t + 1$. Furthermore, by Riemann's theorem, we have (i)

$$\ell(\mathbf{A}) \geq d(\mathbf{A}) + 1 - g = b + 1 - g \geq 1 + t > t.$$

The parts (ii) and (iii) are obvious.   □

6.3.3. *Syndrome Table*

From now on, we shall assume that the list of relations is satisfied. We may have a divisor $\mathbf{U}$ for the syndrome calculation and another divisor $\mathbf{Y}$ for the following syndrome table (which is different from *syndrome calculation*). We assume that $\mathbf{U} = \mathbf{Y}$. Let $\{\phi_0, \ldots, \phi_{u-1}\}$ be a basis for $L(\mathbf{X})$, let $\{\psi_0, \ldots, \psi_{v-1}\}$ be a basis for $L(\mathbf{A})$, and let $\{\chi_0, \ldots, \chi_{w-1}\}$ be a basis for $L(\mathbf{Y})$. Note that $v = \ell(\mathbf{A}) > t$ by the condition (4) on the list. Without losing generality, we use the notation that $S_{ij} = \psi_i \chi_j$ and

$$S_{i,j} = S_{ij} \cdot e = \psi_i \chi_j \cdot e$$

for the error word e. Note that we let the received word be $r = c + e$ with c the original code word and e the error vector, then we have $\mathbf{D} \succ \mathbf{A} + \mathbf{Y} \succ \mathbf{X}$ and $S_{ij} = \psi_i \chi_j \in L(\mathbf{A} + \mathbf{Y}) \subset L(\mathbf{D})$. Therefore, we have $S_{ij} \cdot c = 0$ and

$$S_{i,j} = S_{ij} \cdot e = S_{ij} \cdot r - S_{ij} \cdot c = S_{ij} \cdot r.$$

In other words, although e is to be found, we know $S_{ij} \cdot e$ by computing $S_{ij} \cdot r$. We form a $v \times w$ matrix $[S_{i,j}]$ and call it the *syndrome table*.

6.3.4. Construction of Error Locators

The main reference is Proposition 6.3. Now, we construct an error locator using linear algebra with the help of the preceding propositions. Note that the following proposition is useful not only for **SV Algorithm** but also a good inspiration for **DU Algorithm** of the next section.

Proposition 6.7. *Assume that* $1 \leq wt(e) \leq t$. *Further assume that* $d(\mathbf{Y}) \geq t + 2g - 1$. *The following system of equations has a non-trivial solution:* $[\alpha'_0, \ldots, \alpha'_{v-1}]$ *where* $v > t$,

$$\sum_{k=0}^{v-1} (S_{kj} \cdot r)\alpha_k = 0, \qquad j \leq w = \dim(\mathbf{L}(\mathbf{Y})), \tag{2}$$

which means in matrix form (recall that $S_{i,j} = S_{ij} \cdot r$*),*

$$\begin{bmatrix} \alpha_0 \cdots \cdots \alpha_{v-1} \end{bmatrix} \cdot \begin{bmatrix} S_{0,0} & \cdots & \cdots & S_{0,w-1} \\ \cdots & \cdots & \cdots & \cdots \\ \cdots & \cdots & \cdots & \cdots \\ S_{v-1,0} & \cdots & \cdots & S_{v-1,w-1} \end{bmatrix} = \begin{bmatrix} 0 \cdots \cdots 0 \end{bmatrix}.$$

The left null space of the above matrix is not trivial. For any non-trivial solution $[\alpha'_0, \ldots, \alpha'_{v-1}]$, *let* $\theta = \sum \alpha'_j \psi_j$. *Then,* θ *is an* **error locator** *for* e *in* $L(\mathbf{A})$.

Proof. First thing we want to show is that the above system of equations has a non-trial solution. We have that $\ell(\mathbf{A}) = v > t$. Let M be the set of error locations of e (although we do not know e, the set M exists theoretically). Let us consider the following system of equations:

$$\sum_j \psi_j(P_k)\alpha_j = 0, \qquad P_k \in M.$$

There are t equations in v variables α_i and $v > t$. Hence, there must be a non-trivial solution $[\alpha'_0, \ldots, \alpha'_{v-1}]$. Let $\theta = \sum \psi_j \alpha'_j$. Then, $\theta(P_k) = 0$ for all $P_k \in M$. Therefore, θ is an error locator for e, and $\theta \odot e = 0$.

The only thing we want to verify is that the set $\{\alpha'_0, \ldots, \alpha'_{v-1}\}$ satisfies the above equation (2). It follows from Proposition 6.4 that $\theta \chi \cdot e = 0$ for all

$\chi \in L(\mathbf{Y})$, especially, for all $\chi_j (\in L(\mathbf{Y}))$. Henceforth we have,

$$\sum_k S_{kj} \cdot r\alpha'_k = \sum_k S_{kj} \cdot e\alpha'_k = \sum_k \psi_k \chi_j \cdot ea'_k = \chi_j \cdot (\theta \odot e) = 0$$

for all j. Thus, θ is an error locator for e in $L(\mathbf{A})$. Therefore, equation (2) is satisfied by $\{\alpha'_0, \ldots, \alpha'_{v-1}\}$.

Conversely, let the set $\{\alpha'_0, \ldots, \alpha'_{v-1}\}$ satisfy the above equation (2). Let $\theta = \sum \psi_j \alpha'_j$. It follows from Proposition 6.4 that it suffices to show that $\theta\chi \cdot e = 0$ for all $\chi \in L(\mathbf{Y})$, especially, for all $\chi_j \in L(\mathbf{Y})$. We have

$$\theta\chi_j \cdot e = \sum_k \psi_k \chi_j \cdot ea'_k = \sum_k S_{kj} \cdot e\alpha'_k = \sum_k S_{kj} \cdot r\alpha'_k = 0.$$

Thus, any non-trivial solution of equation (2) induces an error locator. □

6.3.5. *Finding the Error Vector e*

The following proposition tells us how to use an error locator θ to find e.

Proposition 6.8. *Assume that $1 \leq wt(e) \leq t$. Let $\theta \in L(\mathbf{A})$ be an error locator (of e) and M' be the set of positions i such that $\theta(P_i) = 0$. Then, we have the following.* (i) *The cardinal number of M', $\|M'\| \leq d(\mathbf{A})$.* (ii) *Recall $\{\phi_0, \ldots, \phi_{u-1}\}$ is a basis of $L(\mathbf{X}) \subset L(\mathbf{D})$. Moreover, given $d(\mathbf{X}) \geq d(\mathbf{A}) + 2g - 1$, if we solve the following system of equations:*

$$\sum_{P_k \in M'} \phi_j(P_k) E_k = \phi_j \cdot r \quad j = 0, \ldots, u-1, \qquad (3)$$

where the E_k are all indeterminates, then the non-trivial solution exists uniquely. (iii) *Furthermore, the error vector $e = [e_1, \ldots, e_n]$ is a solution set and $e_j = 0$, for all $j \notin M'$.*

Proof. (i) It follows from Proposition 6.6 that since $M' \subset \text{support}(\mathbf{B})$ and $\text{support}(\mathbf{A}) \cap \text{support}(\mathbf{B}) = \emptyset$, all elements of M' are outside the support of $\mathbf{A}$. We have $\|M'\| \leq d(\mathbf{A})$. (ii) Note that $L(\mathbf{X}) \subset L(\mathbf{D})$ and $\phi_j \cdot r = \phi_j \cdot e$. We may replace all $\phi_j \cdot r$ in the above equation by $\phi_j \cdot e$. Let $M = $ all error locations of e, we then have $M \subset M'$. We have

$$\sum_{k \in M'} \phi_j(P_k) E_k = \phi_j \cdot e = \phi_j \cdot r.$$

Since $\phi_j(P_k)e_k = 0$ for all $k \notin M \subset M'$, clearly e is a solution of the above equation. Thus, we know the existence of the solution. We wish to show the uniqueness of the solution. Let e', e^* be two non-trivial solutions. Then, their difference $e' - e^*$ satisfies the following system of equations:

$$\phi_j \cdot (e' - e^*) = 0, \quad j = 1, \ldots, u,$$

which means that $(e' - e^*) \in C_p(\mathbf{B}, \mathbf{X})$, which has a minimal distance $d = d(\mathbf{X}) - 2g + 2 \geq d(\mathbf{A}) + 1$ (by the sufficient relation (6)). On the other hand, e', e^* have only zero values outside M' which has a cardinality of at most $d(\mathbf{A})$. We conclude that $e' - e^* = 0$ or $e' = e^*$. Thus, we prove the uniqueness of the solutions. It means that we may solve the system of equations (3) to find the error vector e. (iii) They are obvious from the above discussion. □

The preceding three propositions are the kernel of the decoding process. Note that always we assume $1 \leq wt(e) \leq t$. If the system of equation (2) produced only trivial solution, then our assumption $1 \leq wt(e) \leq t$ is false, i.e., either there is no error or there are more than t errors. The two cases can be separated by the **check procedure** or by using a **check matrix**. Even if it does produce an error vector e, it may be just an accident, and we still have to check $r + e$ to be sure. If $r + e$ is a code word, then we decode successively. Otherwise, if $r + e$ is not a code word, then the decoder fails.

The above system of equations (3) can be written in matrix form with $i_k \in M'$ as

$$\begin{bmatrix} \phi_0(P_{i_0}) & \cdots & \cdots & \phi_0(P_{i_m}) \\ \cdots & \cdots & \cdots & \cdots \\ \cdots & \cdots & \cdots & \cdots \\ \phi_{u-1}(P_{i_0}) & \cdots & \cdots & \phi_{u-1}(P_{i_m}) \end{bmatrix} \cdot \begin{bmatrix} E_{i_1} \\ \cdot \\ \cdot \\ E_{i_m} \end{bmatrix} = \begin{bmatrix} \phi_0 \cdot r \\ \cdot \\ \cdot \\ \phi_{u-1} \cdot r \end{bmatrix}. \quad (4)$$

Let $E_{i_1} = e_{i_1}, \ldots, E_{i_m} = e_{i_m}$ be a set of solutions of the preceding equation. Since we assume that $1 \leq wt(e) \leq t$ always, i.e., there are at most t errors, which may not be true in the real situation. At this stage, we have to further check if $c = r - e$ is a code word. We may use the *check matrix* or the following procedure. Let $\ell = \ell(\mathbf{D})$ and $\{\phi_0, \ldots, \phi_{u-1}, \ldots, \phi_{\ell-1}\}$ be a basis of $L(\mathbf{D})$. Let us consider the following matrix equation:

$$\begin{bmatrix} \phi_u(\mathbf{P}_{i_1}) & \cdots & \cdots & \phi_u(\mathbf{P}_{i_m}) \\ \cdots & \cdots & \cdots & \cdots \\ \cdots & \cdots & \cdots & \cdots \\ \phi_{\ell-1}(\mathbf{P}_{i_1}) & \cdots & \cdots & \phi_{\ell-1}(\mathbf{P}_{i_m}) \end{bmatrix} \cdot \begin{bmatrix} e_{i_1} \\ \cdot \\ \cdot \\ e_{i_m} \end{bmatrix} = \begin{bmatrix} \phi_u \cdot r \\ \cdot \\ \cdot \\ \phi_{\ell-1} \cdot r \end{bmatrix}. \quad (5)$$

If the above matrix equation is satisfied, then $\phi_j \cdot c = \phi_j \cdot (r - e) = 0$, for all $j = 0, \ldots, \ell - 1$, and c is a code word. Otherwise, the decoding procedure fails, and there are more than t errors.

6.3.6. *Summary of SV Algorithm*

Let C be a smooth projective curve of genus g over a finite field F_q, where $q = 2^m$. We assume that we have *primary Goppa code* (cf. Definition 5.5) $C_p(\mathbf{B}, \mathbf{D})$ which is equivalent to $C_\Omega(\mathbf{B}, \mathbf{D})$ (cf. Proposition 5.7). Thus, we have an integer n. We shall assume that $n > d(\mathbf{D}) > 0$ and $d(\mathbf{D}) > g - 1$; thus, $\ell(\mathbf{D} - \mathbf{B}) = 0$ and $k = n - d(\mathbf{D}) + g - 1$ (cf Proposition 5.2). Let the *minimal distance* $d = d(\mathbf{D}) - 2g + 2$. Let $t = \lfloor \frac{d-g-1}{2} \rfloor$ be the maximal number of errors to be corrected (cf. the remark after Proposition 1.25). Let us select auxiliary divisors $\mathbf{U}$, $\mathbf{A}$, $\mathbf{Y}$, $\mathbf{X}$. The relation between the divisors are $\mathbf{D} \succ \mathbf{U}$, $\mathbf{D} \succ \mathbf{Y} + \mathbf{A} \succ \mathbf{X}$ and $\ell(\mathbf{U}) \geq 2g + t - 1$, $\ell(\mathbf{A}) > t$, $d(\mathbf{A}) < d(\mathbf{D}) - (2g - 2) - t$, $d(\mathbf{X}) \geq d(\mathbf{A}) + 2g - 1, d(\mathbf{Y}) \geq t + 2g - 1$, $\text{support}(\mathbf{Y}) \cap \text{support}(\mathbf{B}) = \emptyset$.

Our consideration is on **a one-point code of the special kind**, we may select $m \geq 3g + 2t - 1$ and b with $t + g \leq b \leq m - 2g - t + 1$ (according to the lemma in Section 6.3, such b exists). Let $\mathbf{P}$ be a rational point disjoint from $\mathbf{B}$, $\mathbf{D} = m\mathbf{P}$, $\mathbf{A} = b\mathbf{P}$, $\mathbf{U} = \mathbf{Y} = (m - b)\mathbf{P}$, $\mathbf{X} = \mathbf{D}$. It is easy to see that all numerical inequalities of this section are satisfied. This SV algorithm can decode up to $\lfloor \frac{d-g-1}{2} \rfloor$ errors, with a speed depending on the speed of solving a system of linear equations.

If we receive a word r, we use the agreed **check matrix** or the **check** procedure to test if r is a code word. If it is a code word, then we pass to the next block. If it is not a code word, then using $L(\mathbf{U})$, we go through the *syndrome calculation* (Proposition 6.5). Or we may go to *syndrome calculation* directly without using the **check matrix** or the **check** procedure first. If the *syndromes* with respect to $L(\mathbf{U})$ are 0's, then either r is a code word or there are more than t errors. A further test using the **check procedure** or the **check matrix** on the received word r will distinguish those two situations. If the *syndromes* with respect to $L(\mathbf{U})$ are not all 0's, solve the system of equations (2) of Proposition 6.7 to find an error locator θ. Using the error locator θ to determine the possible set of errors M' (cf. Proposition 6.6) (which may have more than all error locations of e while with a cardinality $\leq d(\mathbf{A})$). Then, we solve the system of equations (3) of Proposition 6.8 to find the error vector e. At this stage, we have to further check if $c = r - e$ is a code word. We may use the

check matrix or let $\ell = \ell(\mathbf{D})$ and $\{\phi_0, \ldots, \phi_u, \ldots, \phi_{\ell-1}\}$ be a basis of $L(\mathbf{D})$. Let us consider the following matrix equation (from equation (5)):

$$
\begin{bmatrix}
\phi_u(\mathbf{P}_{i_1}) & \cdots & \cdots & \phi_u(\mathbf{P}_{i_m}) \\
\cdots & \cdots & \cdots & \cdots \\
\cdots & \cdots & \cdots & \cdots \\
\phi_{\ell-1}(\mathbf{P}_{i_1}) & \cdots & \cdots & \phi_{\ell-1}(\mathbf{P}_{i_m})
\end{bmatrix}
\cdot
\begin{bmatrix}
e_{i_1} \\
\cdot \\
\cdot \\
e_{i_m}
\end{bmatrix}
=
\begin{bmatrix}
\phi_u \cdot r \\
\cdot \\
\cdot \\
\phi_{\ell-1} \cdot r
\end{bmatrix}.
$$

If the above matrix equation is satisfied, then $\phi_i \cdot c = \phi_i \cdot (r - e) = 0$ for $i = u, \ldots, \ell - 1$. Furthermore, $\phi_i \cdot c = \phi_i \cdot (r - e) = 0$ for $i = 1, \ldots, u$. Therefore, c is a code word for all i, and c is a code word. Otherwise, the **decoding procedure** fails, and there are more than t errors.

Let us consider the following example.

Example 1: Let us consider the Klein quartics projective curve over $\mathbf{F}_{2^4}$ defined by the equation $x^3 y + y^3 + x = 0$. We shall consider **one-point code** with genus $g = 3$. Let us take $t = 3$, $b = 6$, $m = 14 = d(\mathbf{D})$. Then, all numerical conditions are satisfied. This code is with $n = 16, k = 4 = n - d(\mathbf{D}) + g - 1, d = d(\mathbf{D}) - 2g + 2 = 10$.

Pre-computation: All pre-computations are carried out before decoding; hence, they will not be counted in computing time.

(1) Let us count the number of rational points. There are 17 rational points $\{\mathbf{P}_1, \mathbf{P}_2, \ldots, \mathbf{P}_{16}, \mathbf{P}\}$ over $\mathbf{F}_{2^4}$. (cf. **Example 36** in Section 4.9. We shall use the notations of that example). We take $\mathbf{B} = \mathbf{P}_1 + \cdots + \mathbf{P}_{16}$ and $\mathbf{D} = 14\mathbf{P}$, where $\mathbf{P} = $ the origin.

(2) We take $\mathbf{A} = 6\mathbf{P}$, $\mathbf{U} = \mathbf{Y} = 8\mathbf{P}$, $\mathbf{X} = 14\mathbf{P}$. It is easy to check that $d(\mathbf{D}) = 14, d(\mathbf{A}) = 6, d(\mathbf{Y}) = 8, d(\mathbf{X}) = 14 \geq 2 \cdot g - 1 = 5$; therefore, it follows from Proposition 4.28 (Riemann's theorem) that $\ell(\mathbf{D}) = d(\mathbf{D}) + 1 - g, k = n - d(\mathbf{D}) - 1 + g$ and $\mathbf{D} \succ \mathbf{U}, \mathbf{D} \succ \mathbf{A} + \mathbf{Y} \succ \mathbf{X}$.

(3) By direct computation, we know the *rank* of this code is $n - d(\mathbf{D}) + g - 1 = 4$ (cf. Exercises 4 and 5), and the **designed minimal distance** $d \geq d(\mathbf{D}) - 2g + 2 = 10$. The SV algorithm will correct $\lfloor \frac{d-g-1}{2} \rfloor \geq 3$ errors.

(4) We compute a basis of $L(\mathbf{D}) = L(14\mathbf{P})$. It is easy to see that the following $\{f_0, f_3, f_5, f_6, f_7, f_8, f_9, f_{10}, f_{11}, f_{12}, f_{13}, f_{14}\}$ form a basis:

f_0	f_3	f_5	f_6	f_7	f_8	f_9	f_{10}	f_{11}	f_{12}	f_{13}	f_{14}
1	$\frac{1}{x}$	$\frac{y}{x^2}$	$\frac{1}{x^2}$	$\frac{y^2}{x^3}$	$\frac{y}{x^3}$	$\frac{1}{x^3}$	$\frac{y^2}{x^4}$	$\frac{y}{x^4}$	$\frac{1}{x^4}$	$\frac{y^2}{x^5}$	$\frac{y}{x^5}$.

Note that $\text{ord}_{\mathbf{P}}(x) = 3, \text{ord}_{\mathbf{P}}(y) = 1$, and $\text{ord}_{\mathbf{P}}(f_i) = -i$. The reasons that they form a basis are the following: (1) $f_i \in L(14\mathbf{P})$; (2) they are linearly independent over $\mathbf{F}_{2^4}$; (3) by Riemann's theorem, $\ell(14\mathbf{P}) = 14 + 1 - g = 12$. We shall compute the following 12×16 matrix C:

$$
C = \begin{bmatrix}
f_0(\mathbf{P}_1) & \cdots & \cdots & f_0(\mathbf{P}_{16}) \\
f_3(\mathbf{P}_1) & \cdots & \cdots & f_3(\mathbf{P}_{16}) \\
\cdots & \cdots & \cdots & \cdots \\
f_{14}(\mathbf{P}_1) & \cdots & \cdots & f_{14}(\mathbf{P}_{16})
\end{bmatrix}.
$$

It is easy to see that the first 4, f_0, f_3, f_5, f_6, of $\{f_i\}$ form a basis for $L(\mathbf{A}) = L(6\mathbf{P})$ and hence $v = 4$, and the first 6, $f_0, f_3, f_5, f_6, f_7, f_8$, of $\{f_i\}$ form a basis for $L(\mathbf{Y}) = L(\mathbf{U}) = L(8\mathbf{P})$ and hence $w = 6$.

We shall further compute $f_k f_j$ for $f_k \in L(\mathbf{A}) = L(6\mathbf{P})$ and $f_j \in L(\mathbf{Y}) = L(8\mathbf{P})$. We have the following relations:

$$
f_k f_j = f_{k+j} \quad \text{for } (k, j) \neq (5, 7),
$$

$$
f_5 f_7 = f_{12} + f_5 \quad \left(= \left(\frac{y}{x^2} \cdot \frac{y^2}{x^3} = \frac{1}{x^4} + \frac{y}{x^2} \right) \right).
$$

Note that the last equation is equivalent to the defining equation of the curve $x^3 y + y^3 + x = 0$. We compute the *generator matrix* and the **check matrix** which are 16×4 and 4×16 matrices, respectively (cf. Exercises 4 and 5).

6.3.7. *Syndrome Calculation*

Let us consider the received word r, code word c, and error word e. We compute $f_i \cdot r$ for $f_i = \chi_i \in L(\mathbf{Y}) = L(\mathbf{U}) = L(8\mathbf{P})$, i.e.,

$$
\begin{bmatrix}
f_0(\mathbf{P}_1) & \cdots & \cdots & f_0(\mathbf{P}_{16}) \\
f_3(\mathbf{P}_1) & \cdots & \cdots & f_3(\mathbf{P}_{16}) \\
\cdots & \cdots & \cdots & \cdots \\
f_8(\mathbf{P}_1) & \cdots & \cdots & f_8(\mathbf{P}_{16})
\end{bmatrix}
\cdot
\begin{bmatrix}
r_1 \\
r_2 \\
\cdot \\
r_{16}
\end{bmatrix}
=
\begin{bmatrix}
f_0 \cdot r \\
f_3 \cdot r \\
\cdot \\
f_8 \cdot r
\end{bmatrix}.
$$

If $f_i \cdot r = 0$ for $i = 0, 3, 5, 6, 7, 8$, then either there are more than t errors or r is a code word. Only a complete checking procedure can tell the difference. The total number of multiplications is $6 \times 16 = 96$. To finish the checking procedure, We have to compute the following for the remaining

$f_i \in L(D)$:

$$
\begin{bmatrix}
f_9(\mathbf{P}_1) & \cdots & \cdots & f_9(\mathbf{P}_{16}) \\
f_{10}(\mathbf{P}_1) & \cdots & \cdots & f_{10}(\mathbf{P}_{16}) \\
\cdots & \cdots & \cdots & \cdots \\
f_{14}(\mathbf{P}_1) & \cdots & \cdots & f_{14}(\mathbf{P}_{16})
\end{bmatrix}
\cdot
\begin{bmatrix}
r_1 \\ r_2 \\ \cdot \\ r_{16}
\end{bmatrix}
=
\begin{bmatrix}
f_9 \cdot r \\ f_{10} \cdot r \\ \cdot \\ f_{14} \cdot r
\end{bmatrix}.
$$

The total number of computations for the case that there is no error or there is at least one error is $12 \times 16 = 192$.

Note that $f_j = \psi_j \in L(A) = L(6\mathbf{P})$ for $j = 0, 3, 5, 6$. Further, note that $f_k = \varphi_k \in L(X) = L(D) = L(14\mathbf{P})$ for $k = 0, 3, 5, 6, 7, 8, 9, 10, 11, 12, 13, 14$, and $\psi_k \chi_j = f_k f_j$ for $k = 0, 3, 5, 6, j = 0, 3, 5, 6, 7, 8$.

6.3.8. *A Concrete Example of Transmission*

We shall consider a *concrete example*. Let us use the notations of the numerical computations involved in F_{2^4}; let $\alpha, \beta \in \mathbf{F}_{2^4}$ defined by $\alpha^2 + \alpha + 1 = 0, \beta^2 + \beta + \alpha = 0$. It is not hard to see that $\mathbf{F}_{2^2} = \mathbf{F}_2[\alpha]$ and $\mathbf{F}_{2^4} = \mathbf{F}_{2^2}[\beta]$. Let us consider a general number $a\alpha\beta + b\beta + c\alpha + d$, where a, b, c, d are 0 or 1. Let us first represent it as $[abcd]$ a four bits number and then represent the four bits number $[abcd]$ as an integer $a2^3 + b2^2 + c2 + d$ in the *binary* expression. So, all four bits numbers become integers between 0 and 15. Let us order the points $\{\mathbf{P}_i\}$ by the indices. Let us consider a **concrete example**. Certainly, in the real case, we only know the *received word*. However, for the convenience of discussion, let the code word c, error word e, and received word r be as follows:

$$c = 12 \quad 12 \quad 10 \quad 9 \quad 2 \quad 1 \quad 13 \quad 3 \quad 6 \quad 9 \quad 2 \quad 15 \quad 1 \quad 0 \quad 5 \quad 0,$$
$$e = 0 \quad 0 \quad 0 \quad 0 \quad 0 \quad 0 \quad 0 \quad 0 \quad 0 \quad 0 \quad 0 \quad 0 \quad 1 \quad 14 \quad 0 \quad 15,$$
$$r = 12 \quad 12 \quad 10 \quad 9 \quad 2 \quad 1 \quad 13 \quad 3 \quad 6 \quad 9 \quad 2 \quad 15 \quad 0 \quad 14 \quad 5 \quad 15.$$

We use $L(8\mathbf{P})$ for syndrome calculation. The received word r will not pass the *syndrome* calculation, for instance, $f_0 \cdot r = 1 \cdot r$ the sum of r-row as elements in the field which is $\neq 0$.

Finding an Error Locator (cf. Proposition 6.7):

The *syndrome table* $[S_{k,j} = [S_{kj} \cdot r]]$ can be formulated at once as follows:

$k \backslash j$	0	3	5	6	7	8
0	8	11	5	15	2	1
3	11	15	1	3	9	7
5	5	1	9	7	11	6
6	15	3	7	14	6	13

We have to solve the following system of equations to find an error locator θ:

$$\sum_{k=0,3,5,6} S_{k,j}\alpha_k = \sum_{k=0,3,5,6} (S_{kj} \cdot r)\alpha_k = 0, \quad j = 0,3,5,6,7,8.$$

Explicitly, we have

$$8\alpha_0 + 11\alpha_3 + 5\alpha_5 + 15\alpha_6 = 0,$$

$$11\alpha_0 + 15\alpha_3 + 1\alpha_5 + 3\alpha_6 = 0,$$

$$5\alpha_0 + 1\alpha_3 + 9\alpha_5 + 7\alpha_6 = 0,$$

$$15\alpha_0 + 3\alpha_3 + 7\alpha_5 + 14\alpha_6 = 0,$$

$$2\alpha_0 + 9\alpha_3 + 11\alpha_5 + 6\alpha_6 = 0,$$

$$1\alpha_0 + 7\alpha_3 + 6\alpha_5 + 13\alpha_6 = 0.$$

A non-zero solution of the six equations are $\alpha_0 = 3, \alpha_3 = 8, \alpha_5 = 6$, $\alpha_6 = 1$. Note that the multiplication and addition are not the *usual* ones between integers. They are the ones for the field elements. For instance, $2 + 3 = \alpha + \alpha + 1 = 1 \neq 5$ and $2 \times 3 = \alpha \times (\alpha + 1) = 1$ instead of 6. The fastest way of solving the small size system of linear equations is still Gaussian elimination. The number of multiplications involved is $6^2 \times 4/3 = 48$. The corresponding error locator is $\theta = 3f_0 + 8f_3 + 6f_5 + f_6$. Let us find the zero set of θ. We use the following table with the rows of f_0, f_3, f_5, f_6 pre-computed (say, we want to compute $f_5(\mathbf{P}_6)$, we have $f_5(\mathbf{P}_6) = \frac{\alpha\beta + \beta + \alpha + 1}{\alpha^2} = \beta + 1 = 5$), and the values of θ is computed by the formula $\theta = 3f_0 + 8f_3 + 6f_5 + f_6$. We have the following table of values (Table 6.1) to help us do the computation.

To find the zero set, we only have to add the row vectors $[f_j(\mathbf{P}_1), \ldots, f_j(\mathbf{P}_{16})]$ with the produced coefficients and observe the resulting 0 coordinates. For instance, the summation corresponding to $\mathbf{P}_{14}$ is $3 \cdot 1 + 8 \cdot 12 + 6 \cdot 4 + 11$ which written in terms of α, β with $\alpha^2 + \alpha + 1 = 0$,

Table 6.1.

	$P_1,$	$P_2,$	$P_3,$	$P_4,$	$P_5,$	P_6	P_7	P_8	P_9	P_{10}	P_{11}	P_{12}	P_{13}	P_{14}	P_{15}	P_{16}
f_0	1,	1,	1,	1,	1,	1,	1,	1,	1,	1,	1,	1,	1,	1,	1,	1
f_3	0,	0,	3,	2,	3,	3,	2,	2,	14,	13,	10,	8,	15,	12,	9,	11
f_5	0,	0,	1,	1,	4,	5,	6,	7,	12,	15,	9,	11,	5,	4,	7,	6
f_6	0,	0,	2,	3,	2,	2,	3,	3,	8,	10,	14,	13,	9,	11,	12,	15
θ	3,	3,	3,	10,	4,	15,	5,	8,	4,	4,	2,	11,	0,	0,	9,	0

$\beta^2 + \beta + \alpha = 0$ are $(\alpha + 1) + \alpha\beta(\alpha\beta + \beta) + (\alpha + \beta)\beta + (\alpha\beta + \alpha + 1) = \beta^2(\alpha^2 + \alpha + 1) = 0$. In general, it takes $4 \times 16 = 64$ multiplications to find the zero set of θ. So, it is fast. For the present case, the zero set M' is $\{\mathbf{P}_{13}, \mathbf{P}_{14}, \mathbf{P}_{16}\}$.

Finding the Error Vector e (cf. Proposition 6.8):

We have to solve the last set of the first six equations:

$$1E_{13} + 1E_{14} + 1E_{16} = 0,$$

$$15E_{13} + 12E_{14} + 11E_{16} = 8,$$

$$5E_{13} + 4E_{14} + 6E_{16} = 7,$$

$$9E_{13} + 11E_{14} + 15E_{16} = 3,$$

$$10E_{13} + 8E_{14} + 8E_{16} = 2,$$

$$13E_{13} + 13E_{14} + 10E_{16} = 4.$$

The matrix form of the above system of equations is as follows:

$$
\begin{bmatrix}
1 & 1 & 1 \\
15 & 12 & 11 \\
5 & 4 & 6 \\
9 & 11 & 15 \\
10 & 8 & 8 \\
13 & 13 & 10
\end{bmatrix}
\cdot
\begin{bmatrix}
E_{13} \\
E_{14} \\
E_{16}
\end{bmatrix}
=
\begin{bmatrix}
0 \\
8 \\
7 \\
3 \\
2 \\
4
\end{bmatrix}.
$$

According to our Proposition 6.8, there will be an unique non-trivial solution. It is easy to see that $E_{13} = 1$, $E_{14} = 14, E_{16} = 15$ satisfy all equations. For this step, usually it suffices to look at the first three equations. The number of multiplications needed is 9.

6.3.9. *Recovering the Original Message*

We have to further check if the code vector c is really a code vector. For that purpose, note that $f_i \cdot r = \sum f_i(\mathbf{P}_k)r_k = \sum_{k \in M} f_i(\mathbf{P}_k)r_k + \sum_{k \notin M} f_i(\mathbf{P}_k)r_k$ and $f_i \cdot c = \sum f_i(\mathbf{P}_k)c_k = \sum_{k \in M} f_i(\mathbf{P}_k)c_k + \sum_{k \notin M} f_i(\mathbf{P}_k)c_k$, and $e = c - r = c + r$ (since the characteristic is 2), and for $k \notin M$, $f_i(\mathbf{P}_k)r_k = f_i(\mathbf{P}_k)c_k$, and for $k \in M$, $f_i(\mathbf{P}_k)r_k + f_i(\mathbf{P}_k)c_k = f_i(\mathbf{P}_k)e_k$. Further, note that $f_i \cdot c = 0$ (which is what we want to prove!) if and only if we have $\sum_{k \in M} f_i(\mathbf{P}_k)e_k = f_i \cdot r$. Furthermore, for $i = 0, 3, 5, 6, 7, 8$, the equations have been checked, and we have the following system of remaining equations in matrix form:

$$\begin{bmatrix} 10 & 8 & 14 \\ 4 & 5 & 6 \\ 5 & 4 & 7 \\ 11 & 9 & 15 \\ 15 & 12 & 9 \\ 6 & 7 & 5 \end{bmatrix} \cdot \begin{bmatrix} 1 \\ 14 \\ 15 \end{bmatrix} = \begin{bmatrix} 6 \\ 11 \\ 11 \\ 6 \\ 13 \\ 4 \end{bmatrix}.$$

For instance, for the first row, we have

$$10 \cdot 1 + 8 \cdot 14 + 14 \cdot 15$$
$$= (\alpha\beta + \alpha) + (\alpha\beta)(\alpha\beta + \beta + \alpha) + (\alpha\beta + \beta + \alpha)(\alpha\beta + \beta + \alpha + 1)$$
$$= \beta^2 + \alpha^2 + (\alpha\beta + 1)(\beta + \alpha) + \alpha\beta + \alpha\beta + \alpha$$
$$= \beta + \alpha = 6.$$

Indeed, all equations are satisfied which implies that $f_i \cdot c = 0$ for all $0 \le i \le 14$. We conclude that c is a code word, and we decode successfully. Note that if the above equations are not satisfied, then our decoder fails. For this step, it takes 18 multiplications.

The total number of multiplications needed is as follows: (1) If there is no error, (A) we use the *check matrix*, then it is 64; (B) otherwise, we use the *syndrome calculation*, then it is 192. (2) If there are less than t errors, then it is $192 + 48 + 9 + 64 + 18 = 331$. (3) If the decoder fails, then it is $96 + 48 + 9 + 64 + 18 = 235$. We are processing 4 letters which is 16 bits. Per bit, we have (1): (A) 4 multiplications, (B) 12 multiplications; (2) 20.69 multiplications; (3) 14.69 multiplications. The maximal number of multiplications for 16 bits ($k = 4$ and each block contains 4 message symbols) is 32.69 which means 2.04 per bit. A modern computer of 500 mhz can correct at least 15 million blocks or 240 million bits per second. ■

Remark: The main mistake made by some books about the SV algorithm is in the *syndrome* calculation. It is taken to be $S_{kj} \cdot r = \psi_k \chi_j \cdot r$. If all the outcomes are zeroes, then it is faultily claimed that there is no syndrome and no error.

The true *syndrome* calculation uses Proposition 6.4. If there is a received word r that satisfies $\chi \cdot r = 0$ for all $\chi \in L(\mathbf{U})$, where $\mathbf{D} \succ \mathbf{U}$ and $\ell(\mathbf{U}) \ge t + 2g - 1$, then either there are more than t errors or there is no error. The only way to show that there is no error is by using the check matrix or by computing $\phi_j \cdot r = 0$ for a basis $\{\phi_j\}$ of $L(\mathbf{D})$. ■

Exercises

(1) Show that $[1, 1, 1, 2, , 15, 12, , 8, 2, , 8, 6, 14, 11, 0, 1, 0, 0]$ is a code word in **Example 1**.

(2) Show that $[5, 5, 15, 9, 3, 2, 14, 4, 12, 10, 14, 4, 0, 0, 1, 0]$ is a code word in **Example 1**.

(3) Show that $[1, 1, 3, 1, 3, 13, 9, 11, 8, 15, 5, 13, 0, 0, 0, 1]$ is a code word in **Example 1**.

(4) Find a *generator* matrix for the code in **Example 1**.

(5) Find a *check* matrix for the code in **Example 1**.

(6) Write a computer program to decode the code in **Example 1**.

(7) Write down the details of decoding the geometric Goppa code $C_p(\mathbf{B}, 37\mathbf{P})$ with $d(\mathbf{B}) = 64$ based on $x_0^5 + x_1^5 + x_2^5$ with the ground field $\mathbf{F}_{2^4}$ (cf. **Example 35** in Section 4.9).

(8) Write a computer program to decode the code in **Exercise 4**.

6.4. DU Algorithm

Due to the fact that the preceding SV algorithm only decodes up to $\lfloor \frac{d-g-1}{2} \rfloor$ errors instead of the designed number $\lfloor \frac{d-1}{2} \rfloor$ (cf. the **remark** after Proposition 1.25) for a *geometric Goppa* code based on a smooth projective curve C with genus g, and the best *geometric Goppa* codes are based on smooth projective curves with large genus g, note that then $\lfloor \frac{d-g-1}{2} \rfloor \ll \lfloor \frac{d-1}{2} \rfloor$; the shortcoming is serious sometimes. Note that if $g = 0$, there is no difference between $\lfloor \frac{d-g-1}{2} \rfloor$ and $\lfloor \frac{d-1}{2} \rfloor$, and if $g = 1$, then there is small difference. In general, we assume that $g \geq 2$. Feng and Rao found a way to get around it by *majority voting*. Duursma follows their idea and the idea of Weierstrass gaps to extend the SV algorithm to the DU algorithm which will correct errors up to the designed power of $\lfloor \frac{d-1}{2} \rfloor$.

The DU algorithm is more complicated than the SV algorithm.

Summary of DU Algorithm

6.4.1. *Pre-computation*

Let C be a smooth projective curve of genus g ($g \geq 2$) over a finite field F_q, where $q = 2^m$. We assume that we have *primary Goppa code* (cf. Definition 5.5) $C_p(\mathbf{B}, \mathbf{D})$ which is equivalent to $C_\Omega(\mathbf{B}, \mathbf{D})$ (cf. Proposition 5.7). Thus, we have an integer n. We shall assume that $n > d(\mathbf{D}) > 0$ and $d(\mathbf{D}) > g - 1$

and thus $\ell(\mathbf{D} - \mathbf{B}) = 0$ and $k = n - d(\mathbf{D}) + g - 1$ (cf. Proposition 5.2). Let d the *minimal distance* $= d(\mathbf{D}) - 2g + 2$. Let us define $t = \lfloor \frac{d-1}{2} \rfloor$. We have the following **list of relations:** (1) Let $\mathbf{D}$ be a divisor with $d(\mathbf{D}) \geq (2g + 2t - 1)$ (we may take $\mathbf{D} = d(\mathbf{D})\mathbf{P}$). (2) Let $\mathbf{D}' = \mathbf{D} - (d(\mathbf{D}) - 2g - 2t + 1)\mathbf{P}$ (if $\mathbf{D} = d(\mathbf{D})\mathbf{P}$, then $\mathbf{D}' = (2g + 2t - 1)\mathbf{P}$). Then, we have $d(\mathbf{D}') = 2g + 2t - 1$ and $\mathbf{D} \succ \mathbf{D}'$. Therefore, $L(\mathbf{D} - \mathbf{B}) \supset L(\mathbf{D}' - \mathbf{B})$, and every code word in $C_p(\mathbf{B}, \mathbf{D})$ is a code word in $C_p(\mathbf{B}, \mathbf{D}')$. **We decode words in $C_p(\mathbf{B}, \mathbf{D}')$.** One way to think about it is the sender and the receiver should agree on the divisor $\mathbf{D}'$ instead of $\mathbf{D}$ at the beginning. (3) The rank $k = n - d(\mathbf{D}) + g - 1$ (note that according to Proposition 4.28 (Riemann's theorem, if $d(\mathbf{D}) \geq 2g - 1$, then we have $\ell(\mathbf{D}) = d(\mathbf{D}) + 1 - g$, and $k = n - \ell(\mathbf{D}) = n - d(\mathbf{D}) + g - 1$). The minimal distance is $d = d(D') - 2g + 2$, and the *DU* algorithm will correct $t = \lfloor \frac{d-1}{2} \rfloor$ errors (see **remark** in Section 6.6).

Now, we shall choose (4) an auxiliary divisor $\mathbf{A}$ with $d(\mathbf{A}) = t$ and with support of $\mathbf{A}$ disjoint from the support of $\mathbf{B}$ (say, $\mathbf{A} = t\mathbf{P}$).

The secondary auxiliary divisor we use is (5) $\mathbf{A}'$, where $\mathbf{A}'$ is defined by $\mathbf{A}' = \mathbf{D}' - \mathbf{A} - (2g - 1)\mathbf{P}$ (if $\mathbf{A} = t\mathbf{P}$ and $\mathbf{D} = d(\mathbf{D})\mathbf{P}$, then $D' = (2t + 2g - 1)\mathbf{P}$, $\mathbf{A}' = t\mathbf{P} = \mathbf{A}$). Then, $d(\mathbf{A}') = t$. Note that the important divisor $\mathbf{D}' = \mathbf{A} + \mathbf{A}' + (2g - 1)\mathbf{P}$. We shall study $L(\mathbf{A} + i\mathbf{P})$, $L(\mathbf{A}' + i\mathbf{P})$ and $L(\mathbf{A} + \mathbf{A}' + i\mathbf{P})$ for $i = 1, \ldots, 2g - 1$ (see the following).

In this section and the next section, we shall assume that the above list of relations are satisfied. The *syndrome* calculation will not be affected (cf. Section 6.2); we just compute $\phi_i \cdot r$, where ϕ_i runs through a basis of $L(\mathbf{D}') = L(\mathbf{A} + \mathbf{A}' + (2g - 1)\mathbf{P}) \subset L(\mathbf{D})$ and r is the received word. As usual, if the results of the syndrome calculation are zeroes, then either r is a code word or there are more than t errors. We use the *check procedure* or the *check matrix* to tell the difference. The problem is that if the result of the *syndrome calculation* is not zero, then we have to find an error locator θ. Once we find the important error-locator θ, Proposition 6.13 (see below) will lead us to decoding. We shall emphasize to find θ.

6.4.2. *One-Point Code*

Let us consider one-point codes of a special kind.

Let us consider a one-point code $\mathbf{C}_p(\mathbf{B}, m\mathbf{P})$, where $m = 2g + 2t - 1$. Note that for **SV Algorithm**, we need $m \geq 3g + 2t - 1$, i.e., in that case, we need g more numerals for codes for every block. We may simplify the above list of relations to (1) $\mathbf{D} = (2g + 2t - 1)\mathbf{P}$, (2) $\mathbf{D}' = \mathbf{D} = (2g + 2t - 1)\mathbf{P}$, (3) the minimal distance $d = d(\mathbf{D}) - 2g + 2$, (4) $\mathbf{A} = t\mathbf{P}$, and (5) $\mathbf{A}' = \mathbf{A}$.

Under further study, what we really need are only (1) and (4), which we shall use in our study of example. Note that $d = m - 2g + 2 = 2t + 1$ and $t = \frac{d-1}{2} = \lfloor \frac{d-1}{2} \rfloor$, which according to the **remark** after Proposition 1.25 is the best we can do. Comparing with SV algorithm, the value of $m = 2g + 2t - 1$ is much smaller than the value $3g + 2t - 1$ for SV algorithm. The difference of the sizes is g. We have to use the **Feng–Rao's majority voting** to create g lines of values for the decoding purpose.

6.4.3. *Ordering of Basis*

For the convenience of ordering the basis of $L(\mathbf{A} + (2g - 1)\mathbf{P})$ or $L(\mathbf{A'} + (2g - 1)\mathbf{P})$, we shall generalize the classic orderings for Weierstrass gaps. Let $\mathbf{A_0} = \mathbf{A} - t\mathbf{P}$. Then, $d(\mathbf{A_0}) = 0$ and $\mathbf{A} + (2g-1)\mathbf{P} = \mathbf{A_0} + (2g+t-1)\mathbf{P}$. In the following proposition, we study $L(\mathbf{A} + (2g-1)\mathbf{P})$. A similar phenomenon for $L(\mathbf{A'} + (2g - 1)\mathbf{P})$ is claimed verbatim.

Proposition 6.9. *We have* $\ell(\mathbf{A_0} + (2g + t - 1)\mathbf{P}) = g + t$. *We may select a basis* $\{\psi_i\}$ *for* $L(\mathbf{A_0} + (2g + t - 1)\mathbf{P})$, *where* $0 \leq i \leq 2g + t - 1$ *with g of them missing (so, there are precisely $g + t$ of them) such that* $\psi_i \in L(\mathbf{A_0} + i\mathbf{P}) \backslash L(\mathbf{A_0} + (i - 1)\mathbf{P})$, *for* $0 \leq i \leq (2g + t - 1)$.

Proof. It follows from Proposition 4.30 that, since for any canonical divisor $\mathbf{W}$, $\mathbf{W} - (\mathbf{A_0} + (2g + t - 1)\mathbf{P})$ is a divisor of negative degree, then $\ell(\mathbf{W} - (\mathbf{A_0} + (2g+t-1)\mathbf{P})) = 0$. Then, it follows from the Riemann–Roch theorem that $\ell(\mathbf{A_0} + (2g + t - 1)\mathbf{P}) = 2g + t - 1 + 1 - g = g + t$. The other part of the proposition follows from Proposition 4.28 since $\psi_i \in L(\mathbf{A_0} + i\mathbf{P})$ and $\ell(\mathbf{A_0} + i\mathbf{P}) - \ell(\mathbf{A_0} + (i - 1)\mathbf{P}) \leq 1$. $\qquad\square$

It follows from the preceding proposition that we may define as follows.

Definition 6.10. For any rational function $\theta \in L(\mathbf{A_0} + (2g + t - 1)\mathbf{P})$, we define $\nu_{\mathbf{A}}(\theta) = \min\{i : \theta \in L(\mathbf{A_0} + i\mathbf{P})\}$ and call the g missing $\nu_{\mathbf{A}}(\theta)$ the *Weierstrass gaps* of $\mathbf{A}$. We shall call $\nu_{\mathbf{A}}(\theta)$ the *Weierstrass index* of θ. $\blacksquare$

Later, we define the matrix $[S_{i,j}]$ with $S_{i,j} = \psi_i \chi_j \cdot e$, where $i = $ *Weierstrass index of* ψ_i *and* $j = $ *Weierstrass index of* χ_j.

Proposition 6.11. *For a rational function* $\theta \in L(\mathbf{A_0} + (2g+t-1)\mathbf{P})$ *with* $\theta = b_i \psi_i + \sum_{k<i} b_k \psi_k$ *and* $b_i \neq 0$, *we have* $\nu_{\mathbf{A}}(\theta) = i$.

Proof. It follows from the previous proposition. $\qquad\square$

Remark: The classic ordering and the one-point code mentioned above in the subsection are the ones with $\mathbf{A_0} = \mathbf{0}$ and $\nu_\mathbf{A}(\psi_i) = -\operatorname{ord}_\mathbf{P}(\psi_i)$.   ■

Let us write $\mathbf{A} = \sum A(\mathbf{Q}_i)\mathbf{Q}_i$ with $A(\mathbf{Q}_i)$ is the corresponding coefficient of $\mathbf{Q}_i$. We have the following proposition.

Proposition 6.12. *For any rational function $\theta \in L(\mathbf{A_0} + (2g+t-1)\mathbf{P}) = L(\mathbf{A} + (2g-1)\mathbf{P})$, we have $\nu_\mathbf{A}(\theta) = d(\mathbf{A}) - A(\mathbf{P}) - \operatorname{ord}_P(\theta)$ (for notation, see the preceding paragraph) and given $\nu_\mathbf{A}(\theta) = i$, $\nu_{\mathbf{A}'}(\theta') = j$, then $\nu_{\mathbf{A}+\mathbf{A}'}(\theta\theta') = i + j$.*

Proof. Let $\mathbf{A} = \mathbf{E} + s\mathbf{P}$. Then, $\mathbf{A_0} = \mathbf{E} + (s-t)\mathbf{P}$ and we have

$$\theta \in L(\mathbf{A_0} + i\mathbf{P})$$
$$\Leftrightarrow \theta \in L(\mathbf{E} + (i+s-t)\mathbf{P})$$
$$\Leftrightarrow -\operatorname{ord}_P(\theta) \le (i+s-t)$$
$$\Leftrightarrow -\operatorname{ord}_P(\theta) - s + t \le i.$$

Apparently, the minimal possible value for the last inequality is $-\operatorname{ord}_P(\theta) - s + t = d(\mathbf{A}) - \mathbf{A}(P) - \operatorname{ord}_P(\theta)$.

Furthermore, if $\theta \in L(\mathbf{A_0} + i\mathbf{P})\backslash L(\mathbf{A_0} + (i-1)\mathbf{P})$, and $\theta' \in L(\mathbf{A'_0} + j\mathbf{P})\backslash L(\mathbf{A'_0} + (j-1)\mathbf{P})$, then $\theta\theta' \in L(\mathbf{A_0} + \mathbf{A'_0} + (i+j)\mathbf{P})\backslash L(\mathbf{A_0} + \mathbf{A'_0} + (i+j-1)\mathbf{P})$.   □

6.4.4. *Syndrome Table*

Similarly, we shall construct bases $\{\psi_i\}$ for $L(\mathbf{A} + (2g-1)\mathbf{P})$, $\{\chi_i\}$ for $L(\mathbf{A'}+(2g-1)\mathbf{P})$, where $0 \le i \le 2g+t-1$ and $\{\phi_i\}$ for $L(\mathbf{A}+\mathbf{A'}+(3g-1)\mathbf{P})$, where $0 \le i \le 3g + 2t - 1$, respectively. Let $\mathbf{A_0} = \mathbf{A} - t\mathbf{P}$ and $\psi_i \in L(\mathbf{A_0}+i\mathbf{P})\backslash L(\mathbf{A_0}+(i-1)\mathbf{P})$. Note that $\mathbf{D}' = \mathbf{A}+\mathbf{A}'+(2g-1)\mathbf{P}$; therefore, we construct more ϕ_i than needed for coding purposes. Let us consider a special kind of one-point code $\mathbf{A} = \mathbf{A}' = t\mathbf{P}$. Then, we have $\mathbf{A_0} = 0$, $\{\chi_i = \psi_i\}$, $\{\phi_i\}$ basis for $L((2g+t-1)\mathbf{P})$ and $L((3g+2t-1)\mathbf{P})$, respectively. Further, $\psi_i \in L(i\mathbf{P})\backslash L((i-1)\mathbf{P})$, and $\mathbf{D}' = (2g+2t-1)\mathbf{P}$.

We pre-compute the following relations:

$$\psi_i\chi_j = a_{i,j}\phi_{i+j} + \sum_{k<i+j} b_{i,j,k}\phi_k,$$

where $a_{i,j} \ne 0$, without losing generality, we may assume $a_{i,j} = 1$.

Let $\{\phi_i : 0 \leq i \leq (2g + 2t - 1)\} = d(\mathbf{D}')$. We construct the following *syndrome table* which is a $(g + t) \times (g + t)$ matrix S. Customarily, we use the (i, j) entry (where i, j are not Weierstrass gaps) to denote

$$S_{i,j} = \psi_i \chi_j \cdot e.$$

Note that they are not the usual sub-indices, i, j are their Weierstrass indices. We have that in every row and column, there are g indices missing. Since we do not know e, the above matrix only exists theoretically. However, if $0 \leq i, j \leq 2g + t - 1$ and $i + j \leq 2g + 2t - 1$, then $\phi_{i,j} \in L(\mathbf{D}')$ and

$$S_{i,j} = \psi_i \chi_j \cdot e = \psi_i \chi_j \cdot r,$$

which is computable and *known*. For $i + j > 2g + 2t - 1$, the terms $S_{i,j}$ are *unknown*; the reason is that we do not require $\psi_i \chi_j \cdot c = 0$. Hence, $\psi_i \chi_j \cdot e$ may not equal $\psi_i \chi_j \cdot r$, and we cannot use $\psi_i \chi_j \cdot r$ to replace $\psi_i \chi_j \cdot e$. Furthermore, as suggested by the **SV Algorithm**, it would be better to compute $\psi_i \chi_j \cdot e$ for $i + j$ up to $3g + 2t - 1$. Note that $3g + 2t - 1 - (2g + 2t - 1) = g$. What we plan to do is using **Feng–Rao's majority voting method** (see Section 6.6) to find the next g values of $\psi_i \chi_j \cdot e$. Temporarily, we put all $S_{i,j}$, known or unknown, in a matrix form, and it is called the *syndrome table*. It is of the following $(g + t) \times (g + t)$ square form with $*$ known numbers, ? unknown numbers:

$$
\overbrace{\qquad\qquad}^{g+t}
\begin{bmatrix}
* & \cdots & & \cdots & * & * \\
* & \cdots & & \cdots & * & * \\
* & \cdots & & \cdots & * & * \\
* & \cdots & & \cdots & * & * \\
* & \cdots & & \cdots & * & ? \\
* & \cdots & & * & ? & ? \\
* & \cdots & * & ? & ? & ?
\end{bmatrix}.
$$

The indices of the top row (or the leftmost column) of the above table are from 0 to $2g + t - 1$ which are the order of the functions from 0 to $2g + t - 1$. Recall that there are g numbers missing. Therefore, there are $2g + t - 1 + 1 - g = g + t$ terms. So, the above is really a $(g + t) \times (g + t)$ matrix.

6.4.5. *The Construction of Complete Syndrome Table*

To begin with, we know the **syndrome table** $\{S_{i,j}\}$ for all $0 \leq i, j \leq 2g + t - 1$, $i + j \leq 2g + 2t - 1$. So, we know a part of this *syndrome table* bounded by a line $i + j = 2g + 2t - 1$.

In general, we shall build the *syndrome table* line by line, for all i, j with $0 \leq i, j \leq 2g + t - 1$ from the line $i + j = 2g + 2t - 1 + s$ to the next line $i + j = 2g + 2t - 1 + s + 1$ for $s = 0, \ldots, g - 1$. Every time after we construct $S_{i,j}$ (see the next section) for all i, j with $0 \leq i, j \leq 2g + t - 1$, $i + j = 2g + 2t - 1 + s + 1$ for $s = 0, \ldots, g - 1$, we try to find an error locator θ row-wise or column-wise. There are two possibilities: either (1) we find one error locator (then we proceed to decode) or (2) we cannot find one (then, we push to find the $S_{i,j}$ on the next line using the materials of *Feng–Rao's majority voting of the next section*). We shall handle case (1) in this section. Let us start with $s = 0$.

6.4.6. The Construction of Error Locator: Step $s = 0$

As always, we assume that there is an error vector e with $1 \leq wt(e) \leq t$, while completely unknown to us. With this meager information of knowing only the first few rows and columns (for Weierstrass indices (i, j), they are with $\leq t$, $j \leq 2g+t-1$) or $i \leq 2g+t-1$, $j \leq t$ completely, we may consider the first case (do the column-wise first), we **may** be able to find an error locator θ row-wise, i.e., solving the following system of equations for $\{\alpha_i\}$, say, the non-zero solutions form a set $\{\alpha_i'\}$.

$$\sum_{i \leq t} S_{i,j}\alpha_i = 0, \qquad j \leq 2g + t - 1. \tag{2'}$$

Note that $\{S_{i,j}\}$ are known for $i + j \leq 2g + 2t - 1$, the above system of equations has more equations $(2g + t - 1)$ than variables (t). So, in general, there may not be a non-zero solution. We **assume** that there is a non-zero solution, then the non-zero solution $\{\alpha'\}$ can be extended to a solution set of

$$\sum_{i=0}^{2g+t-1} S_{i,j}\alpha_i = 0 \qquad j \leq 2g + t - 1(= w) \tag{2}$$

by assigning the extra $\alpha_i = 0$, where $i \leq 2g + t - 1$.

Let us still call the solution set $\{\alpha_i'\}$. Let $\theta = \sum \alpha_i'\psi_i$. It follows from Proposition 6.4 that θ is an error detector if we replace $\mathbf{Y}$ in that proposition by $\mathbf{A} = \mathbf{A_0} + (t + 2g - 1)\mathbf{P}$. Note that we **assume** that there is a non-zero solution $\{\alpha_i'\}$. Similarly, we may solve the system of equations column-wise to try to find an error locator if we can (if we **cannot** find error locator either ways for all points (i, j) on the line $i + j = 2g + 2t - 1$, then use the materials of the next section to construct the *syndrome table* of the next line $i + j = 2g + 2t - 1 + 1$). If we can find an error locator θ this way (or row-wise), then we may use the following proposition to solve the decoding problem.

Proposition 6.13. *Assume that* $1 \leq wt(e) \leq t$. *Then, we have the following:* (1) *If the above system of equations* $(2')$ *has a non-trivial solution* $\{\alpha'_j\}$ *for* $j \leq t$, *then* $\theta = \sum_{j \leq t} \alpha'_j \psi_j$ *is an error locator. Conversely, if we suppose that there is an error locator* $\theta = \sum_{j \leq t} \alpha'_j \psi_j$, *then the set* $\{\alpha'_j\}$ *is a non-trivial solution of the system of equations* $(2')$. (2) *Suppose that there is an error locator* $\theta \in L(\mathbf{A})$, *where* $\mathbf{A} = \mathbf{A_0} + t\mathbf{P}$. *Let the set* $M' = \{\mathbf{P}_i : \theta(\mathbf{P}_i) = 0\}$. *Then, its cardinality is at most* t, *and the following system of equations*

$$\sum_{\mathbf{P_k} \in M'} \phi_i(\mathbf{P_k})E_k = \phi_i \cdot r(= \phi_i \cdot e), \quad i \leq 2g + 2t - 1 \qquad (3'')$$

has a unique non-zero solution $e \mid_{M'}$, *where* $e \mid_{M'}$ *is the restriction of* e *to the coordinates determined by* M'. *Note that outside the positions determined by* M', e *has coordinates zeroes, so we can find* e *by the preceding information.*

Proof. (1) The first part follows from the discussions before the proposition which show the existence of the error locator θ. Let us prove the second part of (1). We have $\theta \odot e = 0$. Hence, $0 = \chi_j \cdot (\theta \odot e) = \sum_{i \leq t} \alpha'_i \psi_i \cdot (\chi_j \odot e) = \sum_{i \leq t} \alpha'_i \psi_i \chi_j \cdot e = \sum_{i \leq t} \alpha'_i S_{i,j}$.

(2) Since $\theta \in L(\mathbf{A})$, it follows from Proposition 6.6 that the cardinality of $M' \leq d(\mathbf{A}) = t$.

We have

$$\sum_{\mathbf{P_k} \notin M'} \phi_i(\mathbf{P_k})E_k + \sum_{\mathbf{P_k} \in M'} \phi_i(\mathbf{P_k})E_k = \sum_{\mathbf{P_k} \in M'} \phi_i(\mathbf{P_k})E_k$$

since $e_k = 0$ for all $\mathbf{P_k} \notin M'$. Clearly, $e \mid_{M'} = e'$ is a solution of $(3'')$. Let e^* be another solution. Then, the weight of $e' - e^* \leq t$, and $e' - e^*$ will be a solution of the homogeneous system of equations

$$\sum_{\mathbf{P_k} \in M'} \phi_i(\mathbf{P_k})E_k = 0, \qquad i \leq 2g + 2t - 1. \qquad (5')$$

It means that $e' - e^*$ is a code word $\in C_p(\mathbf{B}, \mathbf{A} + \mathbf{A'} + (2g - 1)\mathbf{P})$, while $C_p(\mathbf{B}, \mathbf{A} + \mathbf{A'} + (2g-1)\mathbf{P}))$ has a minimal distance $\geq 2g + 2t - 1 - (2g-2) = 2t + 1 > t \geq (wt(e' - e^*))$. Therefore, $e' - e^* = 0$, and the solution is unique.

It means that we may solve the system of equations $(3'')$ to find the error vector e. □

6.4.7. The Construction of Error Locator: Step $s + 1 < g$

Assume that $1 \leq wt(e) \leq t$. Suppose that by induction, we have the terms $S_{i,j}$ for $i \leq t + s$ and $j \leq 2g + t - 1$. We shall look for an error locator. **Suppose** that we find a non-zero solution $\{\alpha_i'\}$ of the following system of equations (if we **cannot** find error locator either ways for all points (i, j) on the line $i + j = 2g + 2t + s - 1$, then use the materials of the next section to construct the *syndrome table* of the next line $i + j = 2g + 2t + s - 1 + 1$):

$$\sum_{i \leq t+s} S_{i,j}\alpha_i = 0, \qquad j \leq 2g + t - 1. \tag{$2''$}$$

Note that the number of equations is in general greater than the number of variables, and there are no non-zero solutions in general. The non-zero solution $\{\alpha'\}$ can be extended to a solution set of

$$\sum_{i=0}^{v} S_{i,j}\alpha_i = 0 \qquad j \leq w(= 2g + t - 1) \tag{2}$$

by assigning the extra $\alpha_i = 0$, where $v = t + 2g - 1$. Let us still call the solution set $\{\alpha_i'\}$. Let $\theta = \sum \alpha_i'\psi_i$, then θ is an error locator in $L(\mathbf{A_0} + (s + t)\mathbf{P})$ (cf. Proposition 6.4). The following proposition will guarantee that we will be able to decode the message.

Proposition 6.14. *Assume $1 \leq wt(e) \leq t$. Then, we have the following.* (1) *If the above system of equations $(2'')$ has a non-trivial solution $\{\alpha_i'\}$ for $i \leq t + s$, then $\theta = \sum_{j \leq t+s} \alpha_i'\psi_i$ is an error locator. Conversely, if there is an error locator $\theta = \sum_{i \leq t+s} \alpha_i'\psi_i$, then the set $\{\alpha_i'\}$ is a non-trivial solution of the system of equations $(2'')$.* (2) *Let the set $M' = \{\mathbf{P}_i : \theta(\mathbf{P}_i) = 0\}$. Then, its cardinality is at most $s + t$, and the following system of equations*

$$\sum_{\mathbf{P}_k \in M'} \phi_\ell(\mathbf{P}_k)E_k = \phi_\ell \cdot r(= \phi_\ell \cdot e), \quad \ell \leq 2g + 2t + s \tag{$3'$}$$

has a unique non-zero solution $e\mid_{M'}$, where $e\mid_{M'}$ is the restriction of e to the coordinates determined by M'. Note that outside the positions determined by M', e has coordinates zeroes, so we can find e by the preceding information.

Proof. (1) The discussions before the proposition show the existence of the error locator θ. Let us prove the second part of (1). We have $\theta \odot e = 0$. Hence

$$0 = \chi_j \cdot (\theta \odot e) = \sum_{i \leq t+s} \alpha_i'\psi_i \cdot (\chi_j \odot e) = \sum_{i \leq t+s} \alpha_i'\psi_i\chi_j \cdot e \sum_{i \leq t+s} S_{i,j}.$$

(2) Since $S_{i,j} = \psi_i \chi_j \cdot r$ are all known for $0 \leq i, j \leq 2g + t - 1$ and $0 \leq i + j \leq 2g + 2t + s$. Furthermore, by our pre-computation, we have

$$\psi_i \chi_j \cdot r = \phi_{i+j} \cdot r + \sum_{k < i+j} b_{i,j,k} \phi_k \cdot r,$$

then $\phi_\ell \cdot e$ are all known for $\ell \leq 2g + 2t + s$. Since $\theta \in L(\mathbf{A} + s\mathbf{P})$, then it follows from 6.6 proposition that the cardinality of $M' \leq d(\mathbf{A} + s\mathbf{P}) = (s + t)$.

We have

$$\sum_{\mathbf{P}_k \notin M'} \phi_i(\mathbf{P}_k) E_k + \sum_{\mathbf{P}_k \in M'} \phi_i(\mathbf{P}_k) E_k = \sum_{\mathbf{P}_k \in M'} \phi_i(\mathbf{P}_k) E_k$$

since $e_k = 0$ for all $\mathbf{P}_k \notin M'$. Therefore, the above equation $(3')$ has a solution. Clearly $e \mid_{M'} = e'$ is a solution of $(3')$. Let e^* be another solution. Then, the weight of $e' - e^* \leq (s + t + 1)$, and $e' - e^*$ will be a solution of the homogeneous system of equations:

$$\sum_{\mathbf{P}_k \in M'} \phi_i(\mathbf{P}_k) E_k = 0, \quad i \leq 2g + 2t - 1. \tag{5'}$$

It means that, $e' - e^*$ is a code word $\in C_p(\mathbf{B}, \mathbf{A} + \mathbf{A}' + (2g + s)\mathbf{P})$, while $C_p(\mathbf{B}, \mathbf{A} + \mathbf{A}' + (2g+s)\mathbf{P})$ has a minimal distance $\geq 2g + 2t + s - (2g - 2) = 2t + s + 2 > t + s + 1 \geq wt(e' - e^*)$. Therefore, $e' - e^* = 0$, and the solution is unique. It means that we may solve the system of equations $(3')$ to find the error vector e. □

Remark: In Proposition 6.20, we prove the theorem of **majority voting**. We need the statement that there is no error locator θ with $\nu_A(\theta) \leq t + s$, while the above proposition is one of a sequence of propositions that if there is a error locator θ with $\nu_A(\theta) \leq t + s$, then the decoding problem can be solved. After we proved the sequence of propositions to that effect, then the only case which require our attention will have no error locator θ with $\nu_A(\theta) \leq t + s$. Hence, part of the assumptions of Proposition 6.20 is justified. ■

6.4.8. *The Construction of Error Locator:*
Final Step $s + 1 = g$

This is the final step in our discussion of the usage of syndrome table. Assume that $1 \leq wt(e) \leq t$. Although we do not know e, we do know sufficient many $S_{i,j} = \psi_i \chi_j \cdot e$. We first show that there is an error locator θ. Then, we expect to find the error word e.

Due to the restrictions on the indices $0 \leq i, j \leq 2g + t - 1$ and $i + j = 2g + 2t - 1 + (g - 1) + 1 = 3g + 2t - 1$. Therefore, the *syndrome table* can be constructed partially up to the first Weierstrass indices $\leq 2g + t - 1$ (or up to the second indices $\leq 2g + t - 1$), and $i + j \leq 3g + 2t - 1$ and the value $\phi_i \cdot e$ can be found for all $0 \leq i \leq 3g + 2t - 1$.

Let M be the set of error locations of e (although we do not know e, the set M exists theoretically). Let us consider the following system of equations:

$$\sum_{i \leq g+t} \psi_i(\mathbf{P}_k)\alpha_i = 0, \quad \mathbf{P}_k \in M.$$

There are t equations in $t + 1$ variables α_i. Hence, there must be a non-trivial solution $[\alpha_0', \ldots, \alpha_{g+t}']$. Let $\theta = \sum \psi_i \alpha_i'$. Then, $\theta(\mathbf{P}_k) = 0$ for all $\mathbf{P}_k \in M$. **Therefore, θ is an error locator** for e and $\theta \odot e = 0$. The trouble is that we do not know the set M, the above equations only exist in the virtue world, and we are not allowed to solve them.

We shall replace the above equations indexed by the set M by the following system of equations:

$$\sum_{i \leq g+t} S_{i,j}\alpha_i = 0, \qquad j \leq 2g + t - 1. \tag{2$'$}$$

We want to verify that the set $\{\alpha_0', \ldots, \alpha_{g+t}'\}$ satisfies the above equation (2$'$). It follows from Proposition 6.4 that $\theta\chi \cdot e = 0$ for all $\chi \in L(\mathbf{A}' + (2g-1)\mathbf{P})$, especially, it suffices for a basis $\{\chi_j\}$ for $L(\mathbf{A}' + (2g-1)\mathbf{P})$. Henceforth we have,

$$\sum_i S_{ij} \cdot e\alpha_i' = \sum_i \psi_i\chi_j \cdot e\alpha_i' = \chi_j \cdot (\theta \odot e) = 0$$

for all j. Therefore, equations (2$'$) are satisfied.

Conversely, let the non-zero set $\{\alpha_0', \ldots, \alpha_{g+t}'\}$ satisfy the equation (2$'$). Let $\theta = \sum \psi_j \alpha_j'$. It follows from Proposition 6.4 that it suffices to show that $\theta\chi \cdot e = 0$ for all $\chi \in L(\mathbf{A}' + (2g-1)\mathbf{P})$. It suffices to check for a basis $\{\chi_j\}$. We have

$$\theta\chi_j \cdot e = \sum_i \psi_i\chi_j \cdot e\alpha_i' = \sum_i S_{ij} \cdot e\alpha_i' = \sum_i S_{ij} \cdot r\alpha_i' = 0.$$

Thus, any non-trivial solution of equation (2$'$) induces an error locator. It means that the concrete system of equations (2$'$) is equivalent to the virtual equations indexed by the set M for θ.

We have the following proposition.

Proposition 6.15. *Assume that* $1 \leq wt(e) \leq t$. *Let the set* $M' = \{\mathbf{P}_i : \theta(\mathbf{P}_i) = 0\}$. *Then, its cardinality is at most* $g+t$, *and the system of equations*

$$\sum_{\mathbf{P}_k \in M'} \phi_i(\mathbf{P}_k)E_k = \phi_i \cdot e, \quad i \leq 3g + 2t - 1 \qquad (3')$$

has an unique non-zero solution $e \mid_{M'}$, *where* $e \mid_{M'}$ *is the restriction of* e *to the coordinates determined by* M'. *Note that outside the positions determined by* M', e *has coordinates zeroes, so we can find* e *by the preceding information.*

Proof. Since $\theta \in L(\mathbf{A} + g\mathbf{P})$, it follows from 6.6 proposition that the cardinality of $M' \leq d(\mathbf{A} + g\mathbf{P}) = (g + t)$.

Since $c = r + e$ and outside M', $c = r$ and $e = 0$, and the field F_q is of characteristic 2; therefore, we have $\phi \cdot r = \phi \cdot e$ and $\phi \cdot (c + r) = 0$ outside M'. We know that $\phi \cdot r = \phi \cdot e$ for inside M'. Therefore, the above equation $(3')$ has a solution. Clearly, $e \mid_{M'} = e'$ is a solution of $(3')$. Let e^* be another solution. Then, the weight of $e' - e^* \leq (g + t)$ and $e' - e^*$ will be a solution of the homogeneous system of equations

$$\sum_{\mathbf{P}_k \in M'} \phi_i(\mathbf{P}_k)E_k = 0, \quad i \leq 3g + 2t - 1. \qquad (5')$$

It means that $e' - e^* \in C_p(\mathbf{B}, \mathbf{A} + \mathbf{A}' + (3g - 1)\mathbf{P})$, while $C_p(\mathbf{B}, \mathbf{A} + \mathbf{A}' + (3g - 1)\mathbf{P})$ has a minimal distance $\geq 3g + 2t - 1 - (2g - 2) = g + 2t + 1 > g + t \geq wt(e' - e^*)$. Therefore, $e' - e^* = 0$ and the solution is unique. It means that we may solve the system of equations $(3')$ to find the error vector e. $\square$

So, the question is how to construct the *syndrome table* $[S_{i,j}]$ (where $S_{i,j} = S_{ij} \cdot e$) line by line without knowing what is e. The only things which help us are that step by step until the last step, for $s = 0, 1, \ldots, g - 1$, we assume that for all i, j with $i + j \leq 2g + 2t - 1 + s$, $S_{i,j}$ are known, and furthermore, we *assume* that we cannot find an error locator θ with $\nu_{\mathbf{A}}(\theta) \leq t + s$ row-wise (nor column-wise), i.e., the following two system of equations have only trivial solutions:

$$\sum_{i \leq t+s} S_{i,j}\alpha_i = 0, \quad j \leq 2g + t - 1, \qquad (2^*)$$

$$\sum_{j \leq t+s} S_{i,j}\beta_j = 0, \quad i \leq 2g + t - 1. \qquad (2^{**})$$

This fact will be critically important in the proof of Proposition 6.20, which states that with all points $S_{i,j}$ for $i + j \leq 2g + 2t - 1 + s$ classified as *valid votes* and others *invalid votes*, and then the valid votes are further classified as *correct* or *incorrect* votes, the Proposition 6.20 states that the correct vote exists and is the majority of all valid votes. Thus, we shall collect all valid votes, and among the valid votes, we look up the majority block which must be the correct vote. Once the correct vote is found, we make the *vote unanimous* by changing all incorrect *votes*, invalid votes, and non-existent votes to the correct one. Thus, we determined that the values on the line $i + j \leq 2g + 2t - 1 + s$ will be decided correctly.

6.5. Feng–Rao's Majority Voting

As suggested by **SV Algorithm**, it is helpful to know $S_{i,j}$ up to $i + j \leq 3g + 2t - 1$. Let us use the notation $\mathbf{S}|_{i,j}$ to denote the submatrix $[S_{u,v}]$ for $u \leq i, v \leq j$ *and* $i + j \leq 2g + 2t - 1 + s + 1 \leq 3g + 2t - 1$, where $S_{u,v} = \psi_u \chi_v \cdot e$. Now, we study the terms $S_{i,j}$ for $i + j = 2g + 2t - 1 + s + 1$, which means that in the following submatrix, all $*$ terms are *known*, and we want to study the particular unknown term $S_{i,j}$.

$$
S_{|i,j} = \begin{bmatrix} * & \cdots & \cdots & * \\ \cdots & \cdots & \cdots & \cdots \\ * & \cdots & * & * \\ * & \cdots & * & S_{i,j} \end{bmatrix}.
$$

6.5.1. *Construction Process*

We want to find the possible values at $S_{i,j}$. Let us put a variable x at that position. We use Gaussian elimination row-wise (resp. column-wise) to eliminate the last row (resp. column) of $S_{|i,j}$ except at (i, j) position. Depending on the results of these elimination processes, we have a few possibilities as follows: (1) If all numbers at the last row become zero except the last one which is a linear function in x, then we set the value at (i, j) position to be zero, and thus, it determines a value for x. Note that there might be several elimination processes for row operation and even more for column operation. The value of $S_{i,j}$ is determined uniquely (the values determined by row operations and column operations are identical). (2) Otherwise, if it cannot be done, then the constructive way fails, and the term $S_{i,j}$ cannot be determined. We use the following terminologies to clarify the situation. We have the following definition.

Definition 6.16. Let $(S_{i,j})$ be the coefficients matrix of equations (2^*) or (2^{**}) after the proof of Proposition 6.15. If both row-wise and column-wise eliminations mentioned above succeed to determine the value at (i, j) position, the values of $S_{i,j}$ constructed row-wise and column-wise are consistent, i.e., they are equal, the value will be called the *valid vote* of each individual $S_{i,j}$. If the term $S_{i,j}$ cannot be determined by eliminations either row-wise or column-wise, or the solutions in either cases are not unique, or if they both exist and are unique but they do not agree, then we consider that there is no value, and we consider that the *vote* is *invalid*. ∎

6.5.2. *Relations Between All Coefficients of the Matrix* $(S_{i,j})$ *for* $i + j = 2g + 2t - 1 + s + 1$ *with Fixed* s

We shall assume that $S_{i,j}$ are all known for $i + j \leq 2g + 2t - 1 + s$. We are working on the next line $i + j = 2g + 2t - 1 + s + 1$ to find the correct value of $S_{i,j}$ by looking over all possible values.

Let us consider the valid votes. Our task is to find the unknown word e. Before we can determine e, we shall try to find $\psi_i \chi_j \cdot e$. For all $\psi_i \chi_j \cdot e$ in our discussions, we can prove some are the correct numbers, while some are not. We have the following definition.

Definition 6.17. Let us fix s with $s + 1 \leq g$. Let us consider the coefficients matrix $(S_{i,j})$ of equations (2^*) or (2^{**}). If a valid vote $S_{i,j}$ can be shown as $S_{i.j} = \psi_i \cdot \chi_j \cdot e$, then it is called a *correct vote*. ∎

The above definition seems inoperatable without knowing e before hand. The important trick is to show that some $S_{i,j}$ is correct without knowing e. All correct $S_{i,j}$ depending on the index i may be all different; however, there is a way to tell correct one and incorrect one apart, i.e., the correct ones all determine the same $\phi_{i+j} \cdot e$ by the different $\psi_i \chi_j \cdot e$, i.e., let us consider $\{\phi_i : \nu_{\mathbf{A}+\mathbf{A}'}(\phi_i) = i, 0 \leq i \leq 3g + 2t - 1\} \subset L(\mathbf{A} + \mathbf{A}' + (3g - 1)\mathbf{P})$. We may express $\psi_i \chi_j$ for $i + j = 2g + 2t - 1 + s + 1$ (note that $s + 1 \leq g$) as follows:

$$\psi_i \chi_j = a_{i,j} \phi_{i+j} + \sum_{k < i+j} b_{i,j,k} \phi_k. \tag{6}$$

The above is the linear expression of $\psi_i \chi_j$ in terms of a basis ϕ_{i+j} and $\{\phi_k : k < i + j\}$. The coefficients $a_{i,j}, b_{i,j,k}$ can be pre-computed. Moreover, $a_{i,j} \neq 0$; otherwise, $\nu_{\mathbf{A}+\mathbf{A}'}(\psi_i \chi_j) < i + j$. Therefore, since $a_{i,j}$ and all $b_{i,j,k}$

and $\phi_k \cdot e$ are known, then the value $S_{i,j} = \psi_i \chi_j \cdot e$ is uniquely determined by the value of $\phi_{i+j} \cdot e$ and vice versa. We shall consider $S_{i,j}$ as a virtue vote, while by the above formula (6) $\phi_{i+j} \cdot e$ as the common vote by all virtue vote. In this way, we may group all virtue votes in blocks. Later (see Proposition 6.20), we show that it is the block of majority among all blocks the block of correct vote and thus we can tell the correct value of $S_{i,j}$.

A priori, we do not know if there is a correct one. Now, we shall search for all possible values $S_{i,j}$ for fixed i, j with $i + j = 2g + 2t - 1 + s + 1$. Among them we shall determine which ones are valid and which ones form the majority block if there is any.

6.5.3. Possible Values of All Coefficients of the Matrix $(S_{i,j})$ for $i + j = 2g + 2t - 1 + s + 1$ with Fixed s

Let us fix i, j with $i+j = 2g+2t-1+s+1$. In our situation, the matrix $\mathbf{S}|_{i,j}$ is not completely known, i.e., all entries are known except the $S_{i,j}$ term. If the last row of $\mathbf{S}|_{i,j-1}$ is a linear combination of the preceding rows, i.e., the following equations can be solved:

$$\sum_{k<i} \alpha_k S_{k,\ell} = S_{i,\ell} \quad for \ \ell < j, \tag{7}$$

then let $\{\alpha'_0, \ldots, \alpha'_{i-1}\}$ be a solution, then $S_{ij} \cdot e$ can be defined as

$$S_{i,j} = \sum_{k<i} \alpha'_k S_{k,j}.$$

We will have a better understanding of the above procedure by using linear algebra. Let us give $S_{i,j}$ some value to complete the matrix $\mathbf{S}|_{i,j}$.

Proposition 6.18. *Let* $\mathbf{S}|_{i,j} = [S_{k,\ell}]_{(k \leq i, \ell \leq j)}$ *be a submatrix. Then,* (1) $rank(\mathbf{S}|_{i-1,j-1}) = rank(\mathbf{S}|_{i,j})$, $\Leftrightarrow$ $S_{i,j}$ *casts a valid vote;* (2) $rank(\mathbf{S}|_{i-1,j-1}) = rank(\mathbf{S}|_{i-1,j}) = rank(\mathbf{S}|_{i,j-1})$ $\Leftrightarrow$ *there is a unique way to fill the value for* $S_{i,j}$ *so that* $rank(\mathbf{S}|_{i-1,j-1}) = rank(\mathbf{S}|_{i,j})$.

Proof. First, we shall prove claim (1).
($\Rightarrow$) We have

$$rank(\mathbf{S}|_{i-1,j-1}) \leq rank(\mathbf{S}|_{i,j-1}) \leq rank(\mathbf{S}|_{i,j}).$$

So, they must be all equal. Then, by Gaussian elimination to the last row of $\mathbf{S}|_{i,j}$, when the last row of $\mathbf{S}|_{i,j-1}$ becomes 0, the value of $S_{i,j}$ is determined. Hence, our assigned arbitrary value for $S_{i,j}$ must be the determined one.

Suppose that there are two ways to give two different values β_1, β_2 for $S_{i,j}$. If we assign β_2 to the matrix, then clearly the last row of $\mathbf{S}|_{i,j}$ will become $[0, \ldots, 0, a]$ with $a \neq 0$. Then, we have

$$rank(\mathbf{S}|_{i-1,j-1}) = rank(\mathbf{S}|_{i,j}) - 1.$$

It is against our hypothesis. So, the value of $S_{i,j}$ is uniquely determined this way. Similarly, its value is uniquely determined column-wise. Let the value of $S_{i,j}$ be b row-wise and c column-wise. Let us take the value b for $S_{i,j}$ and then apply the column-wise Gaussian elimination to the last column. We get a new matrix with last row $[0, \ldots, 0, b - c]$. It follows that $b = c$.

($\Leftarrow$) If $S_{i,j}$ casts a *valid vote*, then by Gaussian elimination to the last row, the last row will become $[0, \ldots, 0]$. Let us then apply Gaussian elimination to the last column. The last column will be $[0, \ldots, 0]^T$. It is easy to see that

$$rank(\mathbf{S}|_{i-1,j-1}) = rank(\mathbf{S}|_{i,j}).$$

Second, we shall prove claim (2).

Using Gaussian elimination to the last row, the last entry of the last row is 0. Then, the last row consists of 0's only. Since Gaussian elimination is reversible, we determine the value of $S_{i,j}$ uniquely. Now, apply Gaussian elimination to the last column, and we have the last column consisting of 0's only. So, clearly

$$rank(\mathbf{S}|_{i-1,j-1}) = rank(\mathbf{S}|_{i,j}). \qquad \square$$

We want to illustrate the relation of the rank of $\mathbf{S}|_{i,j}$ and the existence of error locators of a certain kind. Let us *assume* that the value of $S_{i,j}$ casts a *valid vote*. Let us consider an error locator θ with $\nu_{\mathbf{A}}(\theta) = i$ and $\theta \in L(\mathbf{A_0} + (2g + t - 1)\mathbf{P})$. Let

$$\theta = a_i \psi_i + \sum_{k<i} b_k \psi_k,$$

where $a_i \neq 0$. As usual, we assume it to be 1. It follows from Proposition 6.4 that we always have

$$\theta \chi_\ell \cdot e = \chi_\ell \cdot (\theta \odot e) = \chi_\ell \cdot 0 = 0, \quad for \ all \ \ell \leq 2g + t - 1.$$

Now, we slightly weaken the conditions in the above equation to the following:

$$\theta \chi_\ell \cdot e = \chi_\ell \cdot (\theta \odot e) = \chi_\ell \cdot 0 = 0, \quad for \ all \ \ell \leq j,$$

or

$$\psi_i \chi_\ell \cdot e + \sum_{k<i} b_k \psi_k \chi_\ell \cdot e = 0, \quad for \ all \ \ell \le j,$$

or simply change the notations as

$$S_{i,\ell} = \sum_{k<i} (-b_k) S_{k,\ell} = 0, \quad for \ all \ \ell \le j.$$

Thus, the last row of $\mathbf{S}|_{i,j}$ is a linear combination of the preceding rows. We rewrite the above in the following way: If the last row of $\mathbf{S}|_{i,j}$ is a linear combination of the preceding rows, i.e., solve the following equations:

$$\sum_{k<i} \alpha_k S_{k,\ell} = S_{i,\ell} \quad for \quad \ell < j. \tag{7}$$

If there is a solution $\{\alpha'_0, \ldots, \alpha'_{i-1}\}$, then $S_{i,j} = S_{ij} \cdot e = \psi_i \chi_j \cdot e$ must be

$$S_{i,j} = \sum_{k<i} \alpha'_k S_{k,j}.$$

Similarly, we do it column-wise. Now, it is clear that the rank of $\mathbf{S}|_{i,j}$ is related to our *voting* procedure.

We shall again explain the above using linear algebra. Let us consider a non-zero vector $v = [\overbrace{\cdots, b_k, \ldots, 1}^{i}, 0, \ldots, 0]$. Then, v is in the left null-space of $S_{|i-1,j} \Leftarrow$ the rational function θ defined by

$$\theta = \psi_i + \sum_{k<i} b_k \psi_k$$

is an error locator with $\nu_{\mathbf{A}}(\theta) = i$ (cf. Proposition 6.4).

We have the following proposition.

Proposition 6.19. *Assume that* $1 \le wt(e) \le t$, *and there is no error locator* $\overline{\theta}, \overline{\theta'}$ *such that* $\nu_{\mathbf{A}}(\overline{\theta}) < i$, $\overline{\theta} \in L(\mathbf{A_0} + (2g+t-1)\mathbf{P})$, *and* $\nu_{\mathbf{A'}}(\overline{\theta'}) < j$, $\overline{\theta'} \in L(\mathbf{A'_0} + (2g+t-1)\mathbf{P})$, *and* $\psi_k \cdot e$ *is correctly defined for* $k < i$, *and* $\chi_k \cdot e$ *is correctly defined for* $k < j$. *Furthermore, we assume that either* (1) *there are error locators* θ *and* θ' *with* $\nu_{\mathbf{A}}(\theta) = i$, $\theta \in L(\mathbf{A_0} + (2g + t - 1)\mathbf{P})$ *and* $\nu_{\mathbf{A'}}(\theta') = j$, $\theta' \in L(\mathbf{A'_0} + (2g + t - 1)\mathbf{P})$ *or* (2) *there is at least one of the two error locators* θ, θ' *and* $S_{i,j}$ *is a valid vote, then* $S_{i,j}$ *casts a vote that is inductively defined and correct. Furthermore, it is obvious that* ϕ_{i+j} *(cf. equation* (6) *after Definition* 6.17) *is inductively defined and correct.*

Proof. Let us assume (1). In the above discussion, we show that equation (7) can be solved and it produces a value for $S_{i,j}$. By Gaussian elimination to the last row, the last row is reduced to $[0, \ldots, 0]$. Now, apply the same reasoning to the columns. The last column is $[0, \ldots, 0]^T$. Therefore, we have $rank(\mathbf{S}|_{i-1,j-1}) = rank(\mathbf{S}|_{i,j})$. It follows from the preceding proposition that $S_{i,j}$ casts a *valid vote*. Let

$$\theta = a_i \psi_i + \sum_{k<i} b_k \psi_k, \tag{1}$$

where $a_i \neq 0$. As usual, factor both sides by a_i, we assume it to be 1. Furthermore, the above expression is unique. Otherwise, take the difference of two expressions, then we have an error locator with $\nu_{\mathbf{A}}$ less than i. Hence, it must be zero, and the expression is unique. We have

$$0 = \theta \odot e = \psi_i \odot e + \sum_{k<i} b_k \psi_k \odot e.$$

It means that we have

$$\psi_i \odot e = \sum_{k<i} b_k \psi_k \odot e, \quad \psi_i \cdot e = \sum_{k<i} b_k \psi_k \cdot e$$

since the characteristic of the field is 2. Now all $\psi_k \cdot e$ are correctly defined as we already know, then $\psi_i \cdot e$ must be correctly defined and unique.

Moreover, if there is an error locator θ' with $\nu_{\mathbf{A}'}(\theta') = j$, $\theta' \in L(\mathbf{A}'_0 + (2g + t - 1)\mathbf{P})$, then repeating the above argument, we conclude that if

$$\theta' = a'_j \chi_j + \sum_{ki'<j} b'_{k'} \chi_{k'},$$

where $a'_j \neq 0$, as usual, we assume it to be 1 and the expression is unique. We have

$$0 = \theta' \odot e = \chi_j \odot e + \sum_{k'<j} b'_{k'} \chi_{k'} \odot e.$$

Since all $\chi_{k'} \odot e$ are correctly defined as we already know, then $\chi_j \odot e$ must be correctly defined and unique. It means that we have

$$\chi_j \odot e = \sum_{k'<j} b'_{k'} \chi_{k'} \odot e, \quad \chi_j \cdot e = \sum_{k'<j} b'_{k'} \chi_{k'} \cdot e.$$

The computation of $\psi_i \chi_j \cdot e$ can be carried out as

$$\psi_i \chi_j \cdot e = \psi_i \cdot (\chi_j \odot e) = \psi_i \cdot \left(\sum_{k' < j} b'_{k'} \chi_{k'} \odot e \right)$$

$$= \sum_{k' < j} \chi_{k'} \cdot (\psi_i \odot b'_{k'} e) = \sum_{k' < j} b'_{k'} \chi_{k'} \cdot (\psi_i \odot e)$$

$$= \sum_{k' < j} b'_{k'} \chi_{k'} \cdot \left(\sum_{k < i} b_k \psi_k \odot e \right)$$

$$= \sum_{k < i, k' < j} b_k b'_{k'} \psi_k \chi_{k'} \cdot e.$$

Although the term e is unknown, we do know $\phi_i \cdot c = 0$, for $i \leq 2g + 2t - 1$. Hence, $\phi_i \cdot e = \phi_i \cdot r$ and are all known, for $i \leq 2g + 2t - 1$. For all h with $2g + 2t - 1 < h \leq 2g + 2t - 1 + s$, inductively, $\phi_h \cdot e$ are known and defined and correct. Then, it follows from the above formula that if the assumption (1) of this proposition are satisfied, it is easy to prove $\psi_i \chi_j \cdot e$ is inductively defined and correct. Furthermore, since

$$\psi_i \cdot \chi_j - \phi_{i+j} = \sum_{k < i+j} b_{i,j,k} \phi_k,$$

it is easy to show that $\phi_{i+j} \cdot e$ is inductively defined and correct.

Let us assume (2). We may assume that θ exists. Let us use the notations above. Then, $\psi_i \odot e$ is correctly defined. Furthermore, we have $\chi_j \theta \cdot e = 0$ and hence we have

$$S_{i,j} = \chi_j \psi_i \cdot e = \chi_j \cdot (\psi_i \odot e).$$

It is easy to see that $S_{i,j}$ is correctly defined. The rest is easy. □

6.5.4. *The Block of Correct Votes is in Majority*

As we point out, we do not know the values of $\phi_\ell \cdot e$ for $2g + 2t - 1 < \ell$. Hence, we cannot test to find $\phi_\ell \cdot e$ directly. We go over all $S_{i,j}$ for $i + j = 2g + 2t - 1 + s + 1$, and collect all *valid votes*. Recall that we assume in the **remark** of the section **The Construction of Error-Locator. Step** $s + 1 < g$, that there are no error locators θ and θ' with $\nu_{\mathbf{A}}(\theta) \leq t + s$, $\theta \in L(\mathbf{A} + (2g-1)\mathbf{P})$ and $\nu_{\mathbf{A}'}(\theta') \leq t+s$, $\theta' \in L(\mathbf{A}' + (2g-1)\mathbf{P})$; otherwise, we can decode (cf. Proposition 6.14). The following proposition says that the majority of valid votes is the correct one.

Proposition 6.20. *Assume that* $1 \leq wt(e) \leq t, 0 \leq s$ *and* $s + 1 \leq g$. *We assume that there are no error locators* θ *and* θ' *with* $\nu_{\mathbf{A}}(\theta) \leq t + s$, $\theta \in L(\mathbf{A} + (2g - 1)\mathbf{P})$ *and* $\nu_{\mathbf{A}'}(\theta') \leq t + s$, $\theta' \in L(\mathbf{A}' + (2g - 1)\mathbf{P})$. *Then, among all valid votes, (the number of correct votes)$-$(the number of incorrect votes)* $\geq s + 1$. *Since* $s \geq 0$, *the correct votes will be in majority. In particular, there is at least one correct vote.*

Proof. Note that $i + j = 2g + 2t - 1 + s + 1$ and $i, j \leq 2g + t - 1$. We conclude that $t + s + 1 \leq i, j \leq 2g + t - 1$.

(1) We claim that there are at least g error locators θ with $\nu_{\mathbf{A}}(\theta)$ distinct numbers between $t + s + 1$ and $2g + t - 1$ and with the corresponding number $a_i = 1$ in the equation (1) of Proposition 6.19.

Since there is no error locator θ with $\nu_{\mathbf{A}}(\theta) \leq t + s$, we may count all error locators in $L(\mathbf{A} + (2g - 1)\mathbf{P})$. Let the divisor $\mathbf{E}$ be $\sum_{\mathbf{P_i} \in M} \mathbf{P_i}$, where M is the set of error locations of e. Clearly, an error locator θ is a non-zero element in $L(\mathbf{A} + (2g - 1)\mathbf{P} - \mathbf{E})$. Note that $d(\mathbf{A} + (2g - 1)\mathbf{P} - \mathbf{E}) \geq 2g - 1$. Therefore, it follows from Riemann's theorem that $\ell(\mathbf{A} + (2g - 1)\mathbf{P} - \mathbf{E}) = g$. Let $\{\theta_i\}$ be a basis. We may use linear changing of basis to make the new basis $\{\theta_i\}$ have distinct values $\nu_{\mathbf{A}}(\theta_i)$ and the corresponding number $a_i = 1$ in the equation (1) of Proposition 6.19. We still name the new basis $\{\theta_i\}$.

(2) Let us count the number m_{cor} of correctly defined *votes* and the number n_{inc} of valid and incorrect *votes*. According to the preceding Proposition 6.19, $m_{cor} \geq$ the number of pairs (i, j) with both error locators θ, θ' with $\nu_{\mathbf{A}}(\theta) = i$, $\nu_{\mathbf{A}'}(\theta') = j$, and $\theta \in L(\mathbf{A} + (2g - 1)\mathbf{P}), \theta' \in L(\mathbf{A}' + (2g - 1)\mathbf{P})$, and $n_{inc} \leq$ the number of pairs (i, j) with no error locators θ with $\nu_{\mathbf{A}}(\theta) = i$, $\theta \in L(\mathbf{A} + (2g - 1)\mathbf{P})$ or error locators θ' with $\nu_{\mathbf{A}'}(\theta') = j$, $\theta' \in L(\mathbf{A}' + (2g - 1)\mathbf{P})$.

Let the set J be the set of all integers $\{i : t + s + 1 \leq i \leq 2g + t - 1\}$, then there is a map π with $\pi(S_{i,j}) = i$ which sends $\{S_{i,j} : i + j = 2g + 2t + s\}$ to J. Let I be the subset of J which is the collection of $\{\nu_{\mathbf{A}}(\theta)\}$ of error locators θ, where the corresponding number $a_i = 1$ in the equation (1) of Proposition 6.19. It means that for every dimension, there is exactly one error locator. So, it follows from (1) that the cardinality of I, $| I |, = g$.

(3) Let I_1 be the subset of J which is the collection of $\{\nu_{\mathbf{A}'}(\theta')\}$ of error locators θ'. Then, by arguments identical to (1) and (2), we have $| I_1 | = g$.

Let us define a reflection σ of the interval $I = \{i : t + s + 1 \leq i \leq 2g + t - 1\}$, i.e., $\sigma(j) = 2g + 2t + s - j$. Note that $\sigma^2 = id$. Then, it is easy to see that $i + j = 2g + 2t + s \Leftrightarrow i = \sigma(j)$ as follows:

$$i + j = 2g + 2t + s \Leftrightarrow i = (2g + 2t + s) - j = \sigma(j).$$

Let us define $I' = \sigma(I_1)$. Then, it is easy to see that $\mid I \mid = g$. Given valid $S_{i,j}$ with $i + j = 2g + 2t + s$ and $i \in I \cup I'$, then we have that either $\theta \in L(\mathbf{A} + (2g - 1)\mathbf{P})$ or $j = \sigma(i), \theta' \in L(\mathbf{A}' + (2g - 1)\mathbf{P})$. In ether case, $S_{i,j}$ is correct. Let the number of incorrect valid vote be m_{inc}. 'We have

$$m_{inc} \leq \mid J \backslash (I \cup I') \mid = 2g - s - 1 - \mid I \cup I' \mid .$$

Moreover, if $i \in I \cap I'$, then $i = \sigma(j)$, and both $\theta' \in L(\mathbf{A}' + (2g - 1)\mathbf{P})$ and $\theta \in L(\mathbf{A} + (2g - 1)\mathbf{P})$ exist. It follows from Proposition 6.19 that $S_{i,j}$ is correct. Let the number of correct valid vote be m_{cor}. We have

$$m_{cor} \geq \mid I \cap I' \mid .$$

So, we have

$$m_{cor} - m_{inc} \geq \mid I \cap I' \mid + \mid I \cup I' \mid + s + 1 - 2g$$
$$= \mid I \mid + \mid I' \mid + s + 1 - 2g > s + 1.$$

So, we proved that more than half of the valid votes are correct; they provide the identical $\theta_{i+j} \cdot e$, and they are in majority. $\qquad \square$

Remark: After we tally all valid *votes* and separate them into blocks according to the values of ϕ_{i+j} induced by them and find the winner, which is the correct one, we make the *vote unanimous* by changing all incorrect *votes*, invalid votes, and non-existent votes to the correct one. Once we complete the extra line $S_{i,j}$ with $i + j = 2g + 2t - 1 + s + 1$, we shall try to find an error locator with order $t + s + 1$ row-wise or column-wise. If we cannot find an error locator either way, then we decide on the next line of $i + j = 2g + 2t - 1 + s + 2$. We thus proceed to the end, using the materials in **The Construction of Complete Syndrome Table, Final Step** $s + 1 = g$, we find the error locator θ and the error vector e. Finally, we solve the decoding problem. $\qquad \blacksquare$

Example 2: Let us consider a one-point code. Let us consider the **Klein** quartics curve over $\mathbf{F_{2^4}}$ defined by the equation $x^3 y + y^3 + x = 0$ (cf. **Example 36** in Section 4.9. We shall use the notation of **Example 36**). Its genus is $g = 3$. Let us consider a one-point code $C_p(\mathbf{B}, \mathbf{D})$ with $\mathbf{B} = \mathbf{P_1} + \cdots + \mathbf{P_{16}}$, and $\mathbf{D} = (2g + 2t - 1)\mathbf{P} = 11\mathbf{P}$ with $t = 3$.

We take $\mathbf{A} = \mathbf{A}' = 3\mathbf{P}$, $A_0 = 0$, and $\mathbf{D}' = \mathbf{D} = 11\mathbf{P}$. For the *geometric Goppa code* $C_p(\mathbf{B}, \mathbf{D})$, we have the rank $k = n - d(\mathbf{D}) + g - 1 = 7$ and the *designed minimal distance* $d = d(\mathbf{D}) - 2g + 2 = 7$. Using the DU algorithm, we expect to correct 3 errors. Comparing with our previous example of **SV Algorithm** we have the following table.

	SV Algorithm	DU Algorithm
$n =$	16	16
$k =$	4	7
$d =$	10	7
$t =$	3	3

Note that all other things of these two algorithms are comparable, but the contents of their messages are different. For a block of length 16, **SV Algorithm** contains information of 4 letters and **DU Algorithm** contains information of 7 letters. Therefore, the same length of transmission of **DU Algorithm** contains $\frac{7}{4}$ amount of information compared with **SV Algorithm**. We shall make the following pre-computation.

(1) We pre-compute the *generator matrix* and the *check matrix* (cf. Exercises 4 and 5).

(2) We compute a basis of $L((3g + 2t - 1)\mathbf{P}) = L(14\mathbf{P}) = L(A + A' + (3g - 1)\mathbf{P})$. It is easy to see that the following: $\{\phi_0, \phi_3, \phi_5, \phi_6, \phi_7, \phi_8, \phi_9, \phi_{10}, \phi_{11}, \phi_{12}, \phi_{13}, \phi_{14}\}$ form a basis.

ϕ_0	ϕ_3	ϕ_5	ϕ_6	ϕ_7	ϕ_8	ϕ_9	ϕ_{10}	ϕ_{11}	ϕ_{12}	ϕ_{13}	ϕ_{14}
1	$\frac{1}{x}$	$\frac{y}{x^2}$	$\frac{1}{x^2}$	$\frac{y^2}{x^3}$	$\frac{y}{x^3}$	$\frac{1}{x^3}$	$\frac{y^2}{x^4}$	$\frac{y}{x^4}$	$\frac{1}{x^4}$	$\frac{y^2}{x^5}$	$\frac{y}{x^5}$,

where $\operatorname{ord}_P(\phi_i) = -i$. The reasons that they form a basis are the following: (1) $\phi_i \in L(14\mathbf{P})$; (2) they are linearly independent over $\mathbf{F}_{2^2}$; (3) by **Riemann's** theorem, $\ell(14\mathbf{P}) = 14 + 1 - g = 12$.

We shall compute the following 12×16 matrix C:

$$
C = \begin{bmatrix}
\phi_0(\mathbf{P_1}) & \cdots & \cdots & \phi_0(\mathbf{P_{16}}) \\
\phi_3(\mathbf{P_1}) & \cdots & \cdots & \phi_3(\mathbf{P_{16}}) \\
\cdots & \cdots \cdots & \cdots \\
\phi_{14}(\mathbf{P_1}) & \cdots & \cdots & \phi_{14}(\mathbf{P_{16}})
\end{bmatrix}.
$$

It is easy to see that the first 6 of the ϕ_i form a basis for $L(\mathbf{A_0} + (2g + t - 1)\mathbf{P}) = L(8\mathbf{P})$).

We shall consider another auxiliary divisor $\mathbf{A} + (2g-1)\mathbf{P} = (2g+t-1)\mathbf{P} = 8\mathbf{P}$. We have to express $\psi_i\chi_j = \sum_{k \le i+j} a_k\phi_k$ as follows (Note that $\chi_j \in L(A' + (2g-1)\mathbf{P}) = L(8\mathbf{P})$ and $\psi_i \in L(A_0 + (2g+t-1)\mathbf{P}) = L(8\mathbf{P})$):

$$\phi_0\phi_i = \phi_i, \quad \text{for } i \le 8$$

$$\phi_3\phi_i = \phi_{i+3}, \quad \text{for } i \le 8$$

$$\phi_5\phi_i = \phi_{i+5}, \quad \text{for } i \le 8, \text{ and } i \ne 7$$

$$\phi_6\phi_i = \phi_{i+6}, \quad \text{for } i \le 8.$$

Using the defining equation $y^3 = x^3y + x$, we deduce the remaining equations as follows:

$$\phi_5\phi_7 = \frac{y^3}{x^5} = \frac{1}{x^4} + \frac{y}{x^2} = \phi_{12} + \phi_5,$$

$$\phi_7\phi_7 = \frac{y^4}{x^6} = \frac{x^3y^2 + xy}{x^6} = \frac{y}{x^5} + \frac{y^2}{x^3} = \phi_{14} + \phi_7.$$

The above are our pre-computations.

6.5.5. *Syndrome Calculation*

Let us consider the received word r. We use the *check matrix* or the **check** procedure to determine if r is a code word. If it is a code word, then we pass to the next block of received message. The computation requires $7 \times 16 = 112$ multiplications. If it is not a code word, then we start the *syndrome calculation* as follows. We compute $\phi_i \cdot r$ for $\phi_i \in L(\mathbf{A}) = L(8\mathbf{P})$, i.e.,

$$\begin{bmatrix} \phi_0(\mathbf{P_1}) & \cdots & \cdots & \phi_0(\mathbf{P_{16}}) \\ \phi_3(\mathbf{P_1}) & \cdots & \cdots & \phi_3(\mathbf{P_{16}}) \\ \cdots & \cdots\cdots & \cdots & \cdots \\ \phi_8(\mathbf{P_1}) & \cdots & \cdots & \phi_8(\mathbf{P_{16}}) \end{bmatrix} \cdot \begin{bmatrix} r_1 \\ r_2 \\ \cdot \\ r_{16} \end{bmatrix} = \begin{bmatrix} \phi_0 \cdot r \\ \phi_3 \cdot r \\ \cdot \\ \phi_8 \cdot r \end{bmatrix}.$$

If $\phi_i \cdot r = 0$ for all ϕ_i, then r is a received word with more than 3 errors (since it fails the check matrix test, it cannot be a code word), we shall return an *error* message and we pass to the next block of the received message. The total number of multiplications is $6 \times 16 = 96$. If $\phi_i \cdot r \ne 0$ for some ϕ_i, we have to compute the following to start the building of our syndrome table:

$$
\begin{bmatrix}
\phi_9(\mathbf{P_1}) & \cdots & \cdots & \phi_9(\mathbf{P_{16}}) \\
\phi_{10}(\mathbf{P_1}) & \cdots & \cdots & \phi_{10}(\mathbf{P_{16}}) \\
\phi_{11}(\mathbf{P_1}) & \cdots & \cdots & \phi_{11}(\mathbf{P_{16}})
\end{bmatrix}
\cdot
\begin{bmatrix}
r_1 \\
r_2 \\
\cdot \\
r_{16}
\end{bmatrix}
=
\begin{bmatrix}
\phi_9 \cdot r \\
\phi_{10} \cdot r \\
\phi_{11} \cdot r
\end{bmatrix} .
$$

The total number of computations for the case that there is an error is $9 \times 16 = 144$.

6.5.6. *A Concrete Example of Transmission*

We use the same representation of F_{2^4} as in **Example 36** of Section 4.9. Recall that we let α be a field generator of $\mathbf{F_{2^2}}$ over $\mathbf{F_2}$, i.e., $\alpha^2 + \alpha + 1 = 0$. Let us consider the field $\mathbf{F_{2^4}}$. Let $\mathbf{F_{2^4}} = \mathbf{F_{2^2}}[\beta]$ with β satisfying $\beta^2 + \beta + \alpha = 0$. It is easy to see that $\{\alpha\beta, \beta, \alpha, 1\}$ is a basis for $\mathbf{F_{2^4}}$ over $\mathbf{F_2}$, i.e., any element $r \in \mathbf{F_{2^4}}$ can be written as $a_0 \cdot 1 + a_1 \cdot \alpha + a_2\beta + a_3 \cdot \alpha\beta$. We represent r as $a_0 + a_1 \cdot 2 + a_2 2^2 + a_3 \cdot 2^3$ as an integer in the following table. Let us consider the example that the code word c, the error word e, and the received word r are as follows:

c =	13	0	14	3	0	10	11	8	1	10	14	12	1	1	1	1,
e =	0	0	0	0	0	0	0	0	0	0	0	0	1	14	0	15,
r =	13	0	14	3	0	10	11	8	1	10	14	12	0	15	1	14.

Certainly the receiver has only the received word r. We include the original message c and the error word e just for our discussions. The whole point is that given only r, we want to recover c.

6.5.7. *Syndrome Tables*

The received word will not pass the *syndrome* calculation. For instance, $\phi_3 \cdot r = \alpha \neq 0$ (we sometimes write $\alpha = 2$ which may cause some confusions). So, we look at the following *syndrome table* (Table 6.2).

Table 6.2 is constructed by the relation $r = c + e$ and $\phi_i\phi_j \cdot r = \phi_i\phi_j \cdot e$ for all $i + j \leq 2g + 2t - 1 = 11$. If there is an error locator θ with $\nu_\mathbf{A}(\theta) \leq 3$ and $\theta \in L(8\mathbf{P})$, then the first two rows must be linearly dependent. Since they are not, there is no such θ (note the requirement of Proposition 6.20). In general, to check if two rows of vectors of length 6 are linearly independent, it takes 6 multiplications at most.

Table 6.2.

	$\phi_0,$	$\phi_3,$	$\phi_5,$	$\phi_6,$	$\phi_7,$	ϕ_8
ϕ_0	0,	8,	4,	3,	12,	8
ϕ_3	8,	3,	8,	7,	11,	11
ϕ_5	4,	8,	11,	11,	$S_{5,7},$	
ϕ_6	3,	7,	11,	$S_{6,6},$		
ϕ_7	12,	11,	$S_{7,5},$			
ϕ_8	8,	11,				

We shall construct for $s = 0$ the terms $S_{i,j}$ for $i + j = 2g + 2t - 1 + 0 + 1 = 12$ (cf. Section 6.6). There are three terms $S_{i,j}$ on the line $i + j = 12$ ($S_{i,j} = \phi_i \phi_j \cdot e$ on the line $i + j = 12$) which are unknown, i.e., $\phi_5 \phi_7 \cdot e = S_{5,7}$, $\phi_6 \phi_6 \cdot e = S_{6,6}$, $\phi_7 \phi_5 \cdot e = S_{7,5}$. Note that the terms $\phi_4 \phi_8 \cdot e = S_{4,8} = \phi_8 \phi_4 \cdot e = S_{8,4}$ are not included in our discussions since ϕ_4 does not exist.

Let us compute the following (3×3) submatrix $\mathbf{S}|_{5,5}$ of Table 6.2:

$$\begin{bmatrix} 0 & 8 & 4 \\ 8 & 3 & 8 \\ 4 & 8 & 11 \end{bmatrix} = \begin{bmatrix} 0 & \alpha\beta & \beta \\ \alpha\beta & \alpha+1 & \alpha\beta \\ \beta & \alpha\beta & \alpha\beta+\alpha+1 \end{bmatrix}.$$

It is easy to see by direct computation that the above matrix is non-singular. It follows from Proposition 6.18 that $S_{5,7}, S_{7,5}$ give only non-valid votes and that the only remaining $S_{6,6} = 7$ must be a correct vote (cf. Propositions 6.18 and 6.20) (hence valid) and by our previous computations, we know that $\phi_{12} \cdot e = \phi_6 \cdot \phi_6 e = 7$, and we shall use the value of $\phi_{12} \cdot e = 7$ to correct the values of $S_{5,7} \cdot e = S_{7,5} \cdot e = \phi_{12} \cdot e + \phi_5 \cdot e$. We have $S_{5,7} = S_{7,5} = \phi_7 \cdot \phi_5 e = \phi_{12} \cdot e + \phi_5 \cdot e = 7 + 4 = \beta + \alpha + 1 + \beta = 3$. In general, to find if $S_{5,7}$ casts a *valid vote*, we have to check if $rank(S_{|3,7}) = rank(S_{|3,6}) = rank(S_{|5,6})$? It needs 13 multiplications. If the answer is yes, then we can find its value with 2 more multiplications and its *vote*. If the answer is no, we discard it. Since the matrix is always symmetric, we do not have to do anything about $S_{7,5}$. For $S_{6,6}$, a similar argument will give us 16 multiplications to determine if the value constitutes a *valid* vote, and if so the *vote*. Therefore, in general we need 31 multiplications to fill those 3 spots.

Table 6.3 is as follows.

Table 6.3.

	ϕ_0,	ϕ_3,	ϕ_5,	ϕ_6,	ϕ_7,	ϕ_8
ϕ_0	0,	8,	4,	3,	12,	8
ϕ_3	8,	3,	8,	7,	11,	11
ϕ_5	4,	8,	11,	11,	3,	$S_{5,8}$
ϕ_6	3,	7,	11,	7,	$S_{6,7}$,	
ϕ_7	12,	11,	3,	$S_{7,6}$,		
ϕ_8	8,	11,	$S_{8,5}$,			

Now, since there is no ϕ_4, then there is no error locator θ with $\nu_{\mathbf{A}}(\theta) = 4$. It means that $\nu_{\mathbf{A}}(\theta) \leq 4 \Leftrightarrow \nu_{\mathbf{A}}(\theta) \leq 3$. From the point of view of linear algebra, there is no $S_{4,8} \cdot e$. Then, we may try to fill Table 6.3 for the next step. We shall try to find the correct values of $S_{5,8}, S_{6,7}, S_{7,6}, S_{8,5}$.

Since the matrix $\mathbf{S}|_{5,5}$ is non-singular, $S_{5,8}, S_{8,5}$ cast invalid votes (cf. Proposition 6.18). We have $s + 1 = 2$, so the number of correct votes must be ≥ 2 (cf. Proposition 6.20) and $S_{6,7}, S_{7,6}$ both cast correct vote. Then, it is routine to find out $S_{6,7} = 12, S_{7,6} = 12$. By correcting all wrong votes and invalid votes, we have $S_{5,8} = 12, S_{8,5} = 12$.

In general, the computations about ranks of various submatrices use $20 + 18 = 30$ multiplications, and to find the values of $S_{5,8}(= S_{8,5}), S_{6,7}(= S_{7,6})$ use $2 + 3 = 5$ multiplications. So, totally it needs 35 multiplications. Table 6.4 becomes as follows.

Table 6.4.

	ϕ_0,	ϕ_3,	ϕ_5,	ϕ_6,	ϕ_7,	ϕ_8
ϕ_0	0,	8,	4,	3,	12,	8
ϕ_3	8,	3,	8,	7,	11,	11
ϕ_5	4,	8,	11,	11,	3,	12
ϕ_6	3,	7,	11,	7,	12,	$S_{6,8}$
ϕ_7	12,	11,	3,	12,	$S_{7,7}$,	
ϕ_8	8,	11,	12,	$S_{8,6}$,		

We check if there is an error locator θ with $\nu_{\mathbf{A}}(\theta) \leq 5$ and $\theta \in L(8\mathbf{P})$, or an error locator θ' with $\nu_{\mathbf{A}'}(\theta') \leq 5$ and $\theta' \in L(8\mathbf{P})$. Now, since the matrix $\mathbf{S}_{|5,5}$ is non-singular, we cannot find any. Therefore, we may use Proposition 6.18.

In general, we have to check if the first three rows are linearly dependent. It takes $12 + 5 = 17$ multiplications.

So, we consider $s = 2, s + 1 = 3$. There are three terms to be considered, $S_{6,8}, S_{7,7}, S_{8,6}$. According to Proposition 6.20, among the valid votes, we have (correct votes)$-$(incorrect votes) $\geq s + 1 = 3$. So, all of them must be valid and correct. On the other hand, $s = 2 = g - 1$ is the end of our induction. So, we should make the last computation.

It can be checked that for the matrix $\mathbf{S}_{|6,7}$, with the rows denoted by $R_0.R_3, R_5, R_6$, we have $R_6 = 6R_5 + 8R_3 + 3R_0$. Therefore, $S_{6,8} = 4 = S_{8,6}$ and $S_{7,7} = \phi_{14} \cdot e + \phi_7 \cdot e = 4 + 12 = \alpha + \alpha\beta + \alpha = 8$.

In general, we have to compute the linear relation between the rows $R_0.R_3, R_5, R_6$ of the matrix $\mathbf{S}_{|6,7}$. Using Gaussian elimination, it takes 26 multiplications, and it takes 3 more multiplications to find the value of $S_{6,8}$. So, totally it takes 29 multiplications to find all $S_{6,8}, S_{7,7}, S_{8,6}$.

We have the following complete Table 6.5.

Table 6.5.

	ϕ_0,	ϕ_3,	ϕ_5,	ϕ_6,	ϕ_7,	ϕ_8
ϕ_0	0,	8,	4,	3,	12,	8
ϕ_3	8,	3,	8,	7,	11,	11
ϕ_5	4,	8,	11,	11,	3,	12
ϕ_6	3,	7,	11,	7,	12,	4
ϕ_7	12,	11,	3,	12,	8,	
ϕ_8	8,	11,	12,	4,		

It follows from Proposition 6.7 that the over-determined linear system of equations $\sum_i S_{i,j}\alpha_i = 0$ for $i = 0, 3, 5, 6, j = 0, 3, 5, 6, 7, 8$. has a non-zero solution $\{\alpha'_i\}$. Then, it follows from Proposition 6.7 that $\theta = \sum \alpha'_i \phi_i$ is an error locator. Now, $\theta = \phi_6 + 6\phi_5 + 8\phi_3 + 3\phi_0$ is an error locator with $\nu_{\mathbf{A}}(\theta) \leq 6$ and $\theta \in L(8\mathbf{P})$. Let us find the zero set of θ. We use the following Table 6.6 with the rows of $\phi_0, \phi_3, \phi_5, \phi_6$ pre-computed and the last row of θ thus computed.

Table 6.6.

	P_1,	P_2,	P_3,	P_4,	P_5,	P_6	P_7	P_8	P_9	P_{10}	P_{11}	P_{12}	P_{13}	P_{14}	P_{15}	P_{16}
ϕ_0	1,	1,	1,	1,	1,	1,	1,	1,	1,	1,	1,	1,	1,	1,	1,	1
ϕ_3	0,	0,	3,	2,	3,	3,	2,	2,	14,	13,	10,	8,	15,	12,	9,	11
ϕ_5	0,	0,	1,	1,	4,	5,	6,	7,	12,	15,	9,	11,	5,	4,	7,	6
ϕ_6	0,	0,	2,	3,	2,	2,	3,	3,	8,	10,	14,	13,	9,	11,	12,	15
θ	3,	3,	3,	10,	11,	13,	9,	15,	8,	3,	13,	9,	0,	0,	9,	0

In general, since $\theta(\mathbf{P}_i) = \phi_6(\mathbf{P}_i) + 6\phi_5(\mathbf{P}_i) + 8\phi_3(\mathbf{P}_i) + 3\phi_0(\mathbf{P}_i)$, we use the pre-computation of $\phi_j(\mathbf{P}_i)$ to locate the zeroes of θ. It takes $4 \times 16 = 64$ multiplications. For the present case, the zero set M' is $\{\mathbf{P}_{13}, \mathbf{P}_{14}, \mathbf{P}_{16}\}$.

Using Proposition 6.13, we have to solve the last set of equations as follows:

$$1E_1 + 1E_2 + 1E_3 = 0,$$
$$15E_1 + 12E_2 + 11E_3 = 8,$$
$$5E_1 + 4E_2 + 6E_3 = 4,$$
$$9E_1 + 11E_2 + 15E_3 = 3,$$
$$10E_1 + 8E_2 + 13E_3 = 12,$$
$$14E_1 + 13E_2 + 10E_3 = 8.$$

According to our Proposition 6.13 , there will be some non-trivial solutions. It is easy to see that $E_1 = 1$, $E_2 = 14$, $E_3 = 15$ satisfies all equations.

In general, there are at most $g + t = 6$ variables, and to solve this system of equations we need 72 multiplications.

Therefore, we conclude

$r = 13 \quad 0 \quad 14 \quad 3 \quad 0 \quad 10 \quad 11 \quad 8 \quad 1 \quad 10 \quad 14 \quad 12 \quad 0 \quad 15 \quad 1 \quad 14,$
$e = 0 \quad 0 \quad 0 \quad 0 \quad 0 \quad 0 \quad 0 \quad 0 \quad 0 \quad 0 \quad 0 \quad 0 \quad 1 \quad 14 \quad 0 \quad 15,$
$c = 13 \quad 0 \quad 14 \quad 3 \quad 0 \quad 10 \quad 11 \quad 8 \quad 1 \quad 10 \quad 14 \quad 12 \quad 1 \quad 1 \quad 1 \quad 1.$

We have to further test to see if c is a code word. We check the following matrix equation:

$$\begin{bmatrix} 13 & 14 & 8 \\ 7 & 6 & 5 \\ 6 & 7 & 4 \end{bmatrix} \cdot \begin{bmatrix} 1 \\ 14 \\ 15 \end{bmatrix} = \begin{bmatrix} 7 \\ 11 \\ 11 \end{bmatrix}.$$

For instance, the first equation is

$$13 \cdot 1 + 14^2 + 8 \cdot 15$$
$$= (\alpha\beta + 1) \cdot 1 + (\alpha\beta + \beta + \alpha)^2 + \alpha\beta(\alpha\beta + \beta + \alpha + 1)$$
$$= \beta + \alpha + 1 = 7.$$

Similarly, we may verify the other two equations. Since the above equation is satisfied, we know that c is a code word and we successively decode. In general, we need 112 multiplications for an error-free block, 407 multiplications to correct 1 block with fewer than three errors and 407 multiplications to return an *error* message (to indicate that there are more than three errors). Since it processes 7 letters which is 28 bits, per bit, it takes 4, 14.5, 14.5 multiplications for error-free, correct errors, and failure, respectively. It is faster than the SV algorithm. ∎

Exercises

(1) Show that $[0, 4, 12, 12, 1, 0, 0, 13, 12, 12, 3, 11, 0, 0, 0, 0]$ is a code word in **Example 2**.

(2) Show that $[13, 9, 9, 4, 0, 14, 1, 0, 13, 4, 9, 6, 0, 0, 0, 0]$ is a code word in **Example 2**.

(3) Show that $[14, 2, 12, 2, 0, 10, 0, 1, 2, 9, 12, 14, 0, 0, 0, 0]$ is a code word in **Example 2**.

(4) Show that $[10, 1, 14, 6, 0, 9, 0, 0, 13, 10, 15, 3, 1, 0, 0, 0]$ is a code word in **Example 2**.

(5) Show that $[0, 6, 3, 9, 0, 5, 0, 0, 6, 0, 5, 3, 0, 1, 0, 0]$ is a code word in **Example 2**.

(6) Show that $[1, 5, 14, 12, 0, 2, 0, 0, 13, 8, 5, 5, 0, 0, 1, 0]$ is a code word in **Example 2**.

(7) Show that $[1, 12, 9, 11, 0, 6, 0, 0, 14, 10, 5, 9, 0, 0, 0, 1]$ is a code word in **Example 2**.

(8) Find a *generator* matrix for the code in **Example 2**.

(9) Find a *check* matrix for the code in **Example 2**.

(10) Write a computer program to decode the code of **Example 2**.

(11) Using DU algorithm, write down the details of decoding the geometric Goppa code $C_p(\mathbf{B}, 37\mathbf{P})$ based on $x_0^5 + x_1^5 + x_2^5$ with the ground field $\mathbf{F}_{2^4}$ (cf. Section 6.8).

Appendices

Appendix A

Convolution Codes

A.1. Representation

The concept of a convolution code was introduced by Elias [19]. Any sequence of data or a stream of data $(a_0, a_1, \ldots, a_n, \ldots)$ with $a_i \in \mathbf{F}$ can be represented as a power series

$$f(x) = \sum_0^\infty a_i x^i.$$

Let us consider the coding process over $\mathbf{F}_2$. Given a stream over $\mathbf{F}_2$, $(a_0, a_1, a_2, a_3, \ldots, a_n, \ldots)$, i.e., $a_i \in \mathbf{F}_2$, we may use the following process to produce two (in general, maybe 1, or any finitely many) data streams $(b_0, b_1, \ldots, b_n, \ldots), (c_0, c_1, \ldots, c_n, \ldots)$.

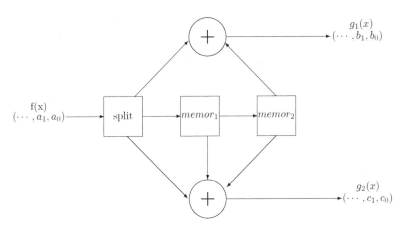

At the beginning, before the count of discrete time $0, 1, 2, \ldots$, we fill the three boxes at the middle row and the two $\oplus$ with zeroes; then, at the time 0, a_0 enters the leftmost box at the middle and all other squares at the middle row and the two $\oplus$ are with 0's. The two $\oplus$ at the top and bottom denote modulo 2 additions. As time goes from n to $n + 1$, all values will move to the next ones according to the arrows, and then the two plus operations in the two $\oplus$ will combine numbers. The $n + 1$th coefficient will occupy the leftmost box. Note that at times $0, 1$, the resulting stream $g_1(x)$ (resp. $g_2(x)$) will receive two zeroes, which we disregard.

Mathematically, let us use

$$g_1(x) = \sum_0^\infty b_i x^i, \quad g_2(x) = \sum_0^\infty c_i x^i,$$

where

$$b_0 = a_0, \; b_1 = a_1, \; b_n = a_n + a_{n-2} \quad \text{for all } n > 1,$$

$$c_0 = a_0, \; c_1 = a_1 + a_0, \; c_n = a_n + a_{n-1} + a_{n-2} \text{ for all } n > 1.$$

Or we simply write

$$g_1(x) = (1 + x^2)f(x),$$

$$g_2(x) = (1 + x + x^2)f(x).$$

It seems that the encoder is simply a multiplication by a polynomial. However, there is a catch: a multiplication by x (a delay of time by 1) should be considered as invertible! So, we should enlarge the polynomial ring $\mathbf{F}[x]$ to the Laurent polynomial ring $\mathbf{F}[x]_{(x)} = \{\frac{h(x)}{x^d} : h(x) \in \mathbf{F}[x]\}$.

If we let the Laurent polynomial ring $\mathbf{F}[x]_{(x)}$ act on the power series ring $\mathbf{F}[[x]]$, then we find it is not even closed under the action induced by x^{-1}. The natural thing to do is to enlarge the power series ring $\mathbf{F}[[x]]$ to the meromorphic function field $\mathbf{F}((x))$. Recall that

$$\mathbf{F}_2((x)) = \left\{ \sum_{-m}^\infty a_i x^i : a_i \in \mathbf{F}_2, m \in \mathbb{Z} \right\}.$$

In the above diagram, the encoded stream $g_1(x)$ is defined by $g_1(x) = (1 + x^2)f(x)$. Mathematically, we need only $g_1(x)$ to determine $f(x)$. Suppose there is no error, mathematically, $f(x) = (1 + x^2)^{-1}g_1(x)$; however, there is a problem: the recursive formula $a_0 = b_0, a_1 = b_1, a_2 = b_2 - b_1, a_3 = b_3 - b_2 + b_1, \ldots$ is getting longer and longer without a bound. Therefore, we require an unbounded number of memory units. Furthermore, if we have a single error for $g_1(x)$, namely replacing $g_1(x)$ by $g_1(x)+x^n$, then the inverse will differ with $f(x)$ at infinitely many places. Any encoder with the last problem will be called a *catastrophic encoder* and will not be used, and we avoid a decoder requiring infinitely many memory units.

A.2. Combining and Splitting

It is clear that if we are allowed only one message series $g_1(x)$ to decode, then the only good encoders which take one incoming stream of data $f(x)$ and produce one stream of data $g_1(x)$ are multiplying by x^n. Those are non-interesting. We should consider using several message series to decode to find one message series. Let us first study a naive technique of combining and splitting data streams. Let

$$f(x) = \sum_{i=-m}^{\infty} a_i x^i$$

$$g_j(x) = \sum_{i=-m}^{\infty} b_{ji} x^i \quad \text{for } j = 1, 2, \ldots, r.$$

Then, we have $h_j(x), h(x)$ uniquely defined by the following equations:

$$f(x) = \sum_{j=0}^{n-1} x^j h_j(x^n),$$

$$h(x) = \sum_{j=0}^{r-1} x^j g_j(x^r).$$

It means that we may split one stream of data $f(x)$ into n streams of data $h_j(x)$ for $j = 0, 1, 2, \ldots, n - 1$, in symbols $S_n(f(x)) = [h_0(x), h_1(x), \ldots, h_{n-1}(x)]$, and combining r streams of data $g_j(x)$ for $j = 0, 1, 2, \ldots, r - 1$ into one stream of data $h(x)$, in symbols $C_r(g_0(x), \ldots, g_{r-1}(x)) = h(x)$. The splitting and combining operations are one to one and onto maps between $\mathbf{F}((x))$ and $\mathbf{F}((x))^r$ or $\mathbf{F}((x))^n$, while they are non-linear respect to the field $\mathbf{F}((x))$.

A.3. Smith Normal Form

The way of splitting a data stream $f(x)$ in in Section A.1 is outside the way described in Section A.2. Let us have a detailed study of it. The splitting can be written mathematically as

$$[f(x)] \begin{bmatrix} 1 + x^2 & 1 + x + x^2 \end{bmatrix} = \begin{bmatrix} g_1(x) & g_2(x) \end{bmatrix}.$$

Using some linear algebra, we may rewrite the above matrix equations as

$$[f(x)] \cdot [1] \cdot [1 \ \ 0] \cdot \begin{bmatrix} 1 + x^2 & 1 + x + x^2 \\ x & 1 + x \end{bmatrix} = \begin{bmatrix} g_1(x) & g_2(x) \end{bmatrix}.$$

It means that we have the following matrix equation:

$$\begin{bmatrix} 1 + x^2 & 1 + x + x^2 \end{bmatrix} = [1] \cdot [1 \ \ 0] \cdot \begin{bmatrix} 1 + x^2 & 1 + x + x^2 \\ x & 1 + x \end{bmatrix}.$$

The above is the well-known *Smith normal form* of a matrix over the P.I.D. $\mathbf{F}_2[x]$ as follows.

Proposition A.1. *Let R be a P.I.D. and M an $r \times n$ matrix with entries in R. Then, M can be written as*

$$M = A\Gamma B,$$

where (1) both A and B are invertible such that their inverses are with entries in R, (2) Γ is in the diagonal form with entries on the diagonal the invariant factors γ_i of M. Note that $\gamma_1 \mid \gamma_2 \mid \ldots \mid \gamma_r$. ∎

Proof. Omitted. □

In our present example, we have $A^{-1} = [1]$ and

$$B^{-1} = \begin{bmatrix} 1 + x & -1 - x - x^2 \\ -x & 1 + x^2 \end{bmatrix}.$$

Therefore, we have the following identity:

$$\begin{bmatrix} g_1(x) & g_2(x) \end{bmatrix} \cdot \begin{bmatrix} 1 + x & -1 - x - x^2 \\ -x & 1 + x^2 \end{bmatrix} \cdot \begin{bmatrix} 1 \\ 0 \end{bmatrix} \cdot [1] = [f(x)].$$

We conclude that with only finitely many memory units (in fact, at most 4), we may recover $f(x)$ if there is no error for $g_1(x)$ and $g_2(x)$. Furthermore, if there are single errors for $g_1(x)$ and $g_2(x)$, say replacing them by x^n, x^m, then the decoding results are polynomials which are not infinitely long meromorphic functions. Therefore, the encoder is not a catastrophic encoder.

In general, we may consider r streams of data $\mathbf{J} = [f_1(x), \ldots, f_r(x)]$ where $f_i(x) \in \mathbf{F}((x))$, we may view it as a vector J in the r-dimensional vector space $\mathbf{F}((x))^r$. An encoder may be viewed as a $r \times n$ matrix M with coefficients in the Laurant polynomial ring $\mathbf{F}[x]_{(x)}$ to produce n streams of data as a vector T in $\mathbf{F}((x))^n$ in the following formula:

$$JM = T.$$

According to the above Smith normal form, we may write $M = A\Gamma B$; the above equation can be rewritten as

$$JA\Gamma B = T.$$

Since both A, B are invertible, Γ is uniquely determined by M. We have Γ to be right invertible $\Longleftrightarrow$ invariant factors of M are invertible, i.e., they are powers of x. Then, and only then, do we have a non-catastrophic encoder.

A.4. Viterbi Algorithm

There are several decoding methods. All are based on maximum-likelihood decoding and are not very difficult. One is the Viterbi algorithm using the *state diagrams*, and the other is the sequential decoding (or Fano algorithm) using the *tree diagram*.

Among all possible decoding methods known today, the Viterbi algorithm is the best. Let us discuss the *Viterbi algorithm*.

Let us consider an example with $f(x) = 1 + x + x^4 + x^6$ as in the representation of Section A.1, then $g_1(x) = (1+x^2)f(x) = 1+x+x^2+x^3+x^4+x^8$, $g_2(x) = (1+x+x^2)f(x) = 1+x^3+x^4+x^5+x^7+x^8$, and $h(x) = g_1(x^2) + xg_2(x^2) = 1+x+x^2+x^4+x^6+x^7+x^8+x^9+x^{11}+x^{15}+x^{16}+x^{17}$, which represents the data stream $[111010111101000111]$. Note that we shall transmit $h(x)$, then we may recover the even part $(g_1(x^2))$ and the odd part $(xg_2(x^2))$, hence $g_1(x)$ and $g_2(x)$, and then $f(x)$. However, a noisy channel produces a received word $r(x) = 1+x+x^2+x^6+x^8+x^9+x^{11}+x^{15}+x^{16}+x^{17}$ which corresponds to $[111000101101000111]$. How do we recover the original $h(x)$? More specifically, how do we recover the original data stream? We shall consider all possible transformations: if the input bit is 1 (i.e., $f(x) = 1 + \cdots$), then the output is $[11]$ (i.e., $h(x) = 1 + x + \cdots$); if the input bit is 0, then the output is $[00]$. If we consider only the possible sequences of coefficients, then it is easy to see that after t steps, we consider 2^t possible paths. The number is explosively large. It is the number of the first t places of binary expansions of all real numbers between 0 and 1. It will become *uncountably infinity* as $n \mapsto \infty$. We shall prune all possible sequences of output coefficient vectors using the Hamming distance to the received word $r(x)$ as the criterion.

Let us consider the following **Figure A.1 of paths** of $1 \leq t \leq 3$.

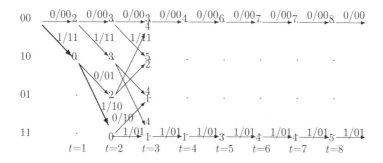

Figure A.1.

We write the *input/output* on most arrow (where the input $=$ the corresponding coefficient of $f(x)$ and the output $=$ the corresponding coefficient of $h(x)$), and write the Hamming distance of the output sequence and the truncated r at the tip of an arrow. For instance, for the path determined by a sequence of inputs $[001]$ ($f(x) = x^2$), the output is $[000011](h(x) = x^4 + x^5 + \cdots)$; comparing with the received sequence $r = [111000 \cdots]$, we find that the Hammong distance is 5. For the path determined by a sequence of input $[110]$ ($f(x) = 1 + x$), the output is $[111010](h(x) = 1 + x + x^2 + x^4 + 0x^5 + \cdots)$; comparing with the received sequence $r = [111000 \cdots](r(x) = 1 + x + x^2 + \cdots)$, we find that the Hammong distance is 1. Looking at the above diagram up to $t = 3$, we realize that since the minimal distance is 1 between all allowed sequences and the received one, there must be some error(s). Furthermore, there are four pairs of paths where for the two paths in the same pair, each one starts and ends at the same digits. Comparing the two paths in each pair, we may delete the one with the larger Hamming distance from the received r. This is the *pruning method*. We shall use the pruning method to keep only four paths for consideration at t (instead of $2^t = 2^3 = 8$ possible paths) and extend the above diagram from $t = 3$ to $t = 8$. We have the following diagram with the details left to the reader. We select the path with Hamming distance 2

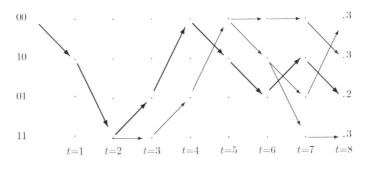

Figure A.2.

to the received word r as the maximum-likelihood decoded result. From the selected path, we decide that the maximum-likelihood code word is [111010111101000111] which accidentally happens to be the original code word (see Figure A.2).

Sphere-Packing Problem and Weight Enumerators

To tile the real plane with same size tiles of same shape, one may use same sized triangles, equilaterals, or hexagons. For all other shapes there are always empty spaces left. This tiling is tight; we may consider the problem of filling a plane with the same size discs with gaps allowed. One knows from experience that the following arrangement for discs is tight, and in fact, it can be proved mathematically.

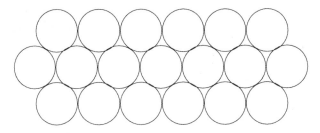

In 1611, Kepler conjectured that the best space-saving way to stack equal sized balls in $\mathbf{R}^3$ was already known to all fruit sellers in the market, the pyramidal arrangements. In 1990, the old open problem was dug up by Hsiang; in fact, he published a controversial paper solving this 400-years-old conjecture (claimed in 1990). Later on, Hales (claimed in 1998) using the *computer* in an essential way solved the problem. For the higher dimensional cases, the only known solutions are for dimensions 4, 8, and 24.

The relation of the sphere-packing problem and the self-correcting code is as follows. Let us take the plane example. If both the sender and the receiver select the centers of the discs as the permissible code words, and the receiver receives a point which is slightly different from the point of the original message, we believe that the distance between those two points indicate the measure of error that occurred. As long as the received point is in a disc (which is likely), the receiver will decode it as the center of the disc. Naturally, we want to pack the space by the most efficient way so we may have the largest possible spheres in some region of the plane, i.e., which corresponds to the largest rate of information. We shall generalize the sphere-packing problem to higher dimension.

B.1. Golay Code

Let us consider the beautiful [23,12] Golay code. In the vector space $V = F_2^{23}$, a ball of radius 3 centered at $[00\cdots00]$ contains at least

$$\sum_{i=0}^{3} C_i^{23} = 2048 = 2^{11}$$

points. It is easy to see any ball of radius 3 contains exactly 2^{11} elements. In the vector space V, there are 2^{23} elements, so it is possible to have 2^{12} balls with radius 3 that do not overlap. Indeed this happens (see the following), and we have a code with 2^{12} code words of length 23 and the information rate $12/23 = 0.52$.

We shall follow the way of constructing the Reed–Solomon code to construct the Golay code. Let us consider the field $F_{2^{11}}$. All non-zero elements in $F_{2^{11}}$ satisfy the following equation:

$$x^{2^{11}-1} + 1 = 0.$$

Note that

$$2^{11} - 1 = 23 \times 89.$$

Therefore, $F_{2^{11}}$ contains a 23*rd* root of unity. We have the following decomposition:

$$x^{23} + 1 = (x + 1)(x^{11} + x^9 + x^7 + x^6 + x^5 + x + 1)$$
$$\times (x^{11} + x^{10} + x^6 + x^5 + x^4 + x^2 + 1) = (x + 1)g(x)g^*(x),$$

where $g^*(x) = x^{11}g(1/x)$. We have the following definition:

Definition B.1. The [23,12] Golay code consists of all polynomials $h(x)g(x)$ in $F_2[x]/(x^{23} - 1)$ with $\deg(h(x)) \leq 11$. ∎

Note that the [23,12] Golay code is not a Reed–Solomon code. We have the following proposition.

Proposition B.2. *It is possible to have 2^{12} balls with radius 3 that do not overlap and pack* $\mathbf{V} = \mathbf{F}_2^{23}$.

Proof. Omitted.   □

The other Golay code is the [11,6] Golay code over $\mathbf{F}_3$, where balls of radius 2 densely pack $\mathbf{F}_3^{11}$.

B.2. Uniformly Packed Code

The perfect codes over a finite field are like the tiling of the plane. The only perfect codes are (1) repetition code of odd length, (2) q-ary Hamming codes (cf. Exercise 1.5 (1), (2), (4)), and (3) [23,12] binary Golay code and [11,6] ternary Golay code. Otherwise, it is impossible to use balls of the same radius to tile a vector space $\mathbf{F}_q^n$. Let us consider the simple case over $\mathbf{F}_2$. We study

$$A(n,d) = \text{the largest integer M such that there exist M codewords}$$
$$\times \{x_1, \ldots, x_M\} \text{ such that } d(x_i, x_j) \geq d, \text{ if } i \neq j.$$

Let d be odd. Then, the number $A(n,d)$ is the maximal number of ways of putting balls of radius $\lfloor d/2 \rfloor$ in $\mathbf{F}_2^n$ without touching each other. There might be some extra space allowed. This is similar to the sphere-packing problem in $\mathbf{R}^3$. We may form a code C with $C = \{x_1, \ldots, x_M\}$ which will correct $\lfloor d/2 \rfloor$ errors. In general, it is difficult to find the number $A(n,d)$ except some easy cases with n, d small or $d \geq n$.

We may put more conditions on the code C to make it easier to construct. Let us define the distance $d(x, C)$ and the covering radius $\rho(C)$ of a set C as

$$d(x, C) = \min\{d(x, c) : c \in C\},$$
$$\rho(C) = \max\{d(x, C) : x \in \mathbf{F}_2^n\},$$
$$d = \min\{d(c_1, c_2) : c_1 \neq c_2 \in C\}.$$

We have the following definition.

Definition B.3. Given a code C which may not be linear. Let e be a positive integer such that the minimal distance d of C satisfies $d \geq 2e + 1$ and $\rho(C) = e + 1$. Then, C is said to be a uniformly packed code with parameter r if every word x with $d(x, C) \geq e$ has distance e or $e + 1$ to exactly r code words. ∎

Let us assume that C is a *uniformly packed code*. Let us consider the condition $d(x, C) \geq e$. If $d(x, C) = e$, then there is an unique $c \in C$ such that $d(x, c) = e$. Otherwise, let $d(x, c_1) = d(x, c_2) = e$ and $c_1 \neq c_2$. Then, it follows from the triangle inequality that $d(c_1, c_2) \leq 2e < d$. A contradiction. Note that in this situation, it is not hard to see that given x each of the r code words $c_1, c_2, \ldots, c_r$ whose sets of coordinates are different from the coordinates of x must be all disjoint; otherwise, since the ground field has only two elements $0, 1$, the distance between two code words c_i with non-disjoint sets of coordinates which are different from the corresponding coordinate of x will be $\leq 2e$ contradicts to $d \geq 2e + 1$. Therefore, we have $(r - 1)(e + 1) + e \leq n$, so we conclude that

$$r - 1 \leq \frac{n - e}{e + 1},$$

$$r \leq \frac{n + 1}{e + 1}.$$

If $d(x, C) = e + 1$, then similarly we have

$$r \leq \frac{n}{e + 1}.$$

We define a *nearly perfect code* as a uniformly packed code with parameter $r = \lfloor \frac{n}{e+1} \rfloor$. After years of research, it turns out that there are few uniformly packed codes and nearly perfect codes (cf. [9], p. 126).

B.3. Weight Enumerators

For any code C, it is meaningful to compute the geometric configuration $\{d(c_i, c_j) : c_i, c_j \in C\}$ of C defined by the Hamming distance. In general, the computation is complicated. We may restrict C to linear $[n, k]$ code only. Then, we may take $c_i = 0$ and define $A_\ell = |\{c_j : c_j \in C, \ wt(c_j) = \ell\}|$. As a standard trick in mathematics, we form the following polynomial, which is called the *weight enumerator* of C:

$$A(z) = A_0 + A_1 z + \cdots + A_n z^n.$$

We have clearly $A_0 = 1$ and $A(1) = q^k$. We have the following example.

Example B1: Let C be the repeated code with generating matrix $[11111]$. Then, $C = \{[00000], [11111]\}$, and $A(z) = 1 + Z^5$. ∎

However, to compute $A(z)$ for large q, we still have to look over all q^k vectors in C. It is still a formidable task. If $n - k$ is small, then we may be able to compute the weight enumerator for the dual code $C^\perp$. The following proposition is courtesy of MacWilliams.

Proposition B.4 (MacWilliams identity). *Let $A(z)$ be the weight enumerator of an $[n, k]$ linear code C and $B(z)$ be the weight enumerator of the dual code $C^\perp$. Then, $A(z)$ and $B(z)$ are related by the following formula:*

$$B(z) = \frac{1}{q^k} \sum_{i=0}^{n} A_i (1 - z)^i (1 + (q - 1)z)^{n-i-1}.$$

Proof. The reader is referred to [6], p. 148 or [9], p. 41. $\qquad\square$

Example B2: Let $C = [7, 4]$ Hamming code. Then, its dual code $C^\perp$ is an $[7, 3]$ code. It turns out that each non-zero code words in $C^\perp$ has weight 4. Thus, the weight enumerator of $C^\perp$ is $1 + 7z^4$. Therefore, the weight enumerator of C which is considered as the dual code of $C^\perp$ is

$$\frac{1}{8}[(1 + z)^7 + 7(1 - z)^4(1 + z)^3] = 1 + 7z^3 + 7z^4 + z^7. \qquad\blacksquare$$

Other Important Coding and Decoding Methods

C.1. Hadamard Codes

The space ship US Mariner 1969 carried a *Hadamard code*. We introduce the *Hadamard matrix* which is defined as follows.

Definition C.1. A square matrix H_n of order n with entries $+1$ or -1 such that $H_n H_n^T = nI$ is called a *Hadamard matrix*. ∎

Example C1: We give the following examples:

$$H_2 = \begin{bmatrix} 1 & 1 \\ 1 & -1 \end{bmatrix} = \begin{bmatrix} + & + \\ + & - \end{bmatrix},$$

$$H_4 = \begin{bmatrix} 1 & 1 & 1 & 1 \\ 1 & 1 & -1 & -1 \\ 1 & -1 & 1 & -1 \\ 1 & -1 & -1 & 1 \end{bmatrix} = \begin{bmatrix} + & + & + & + \\ + & + & - & - \\ + & - & + & - \\ + & - & - & + \end{bmatrix}.$$
∎

It can be shown that there are *Hadamard matrices* only of order 2 or $4m$. In general, given any integer $n = 4m$, we do not know if there is a *Hadamard matrix* of order n. The smallest unknown case is $n = 668$. In H_n and $-H_n$, we replace -1 by 0. Thus, we have $2n$ rows which are code words in $\mathbf{F_2}^n$. It is easy to see that all code words are of equidistant $\frac{n}{2}$. Thus, we have a code, a *Hadamard code*. If n is not a power of 2, then the *Hadamard code* is nonlinear. The best codes with $n \leq 2d$ are practically all nonlinear and related to the *Hadamard matrix*.

C.2. Reed–Muller Code

There are several ways to present the Reed–Muller code which is a binary code correcting several errors. The code was discovered by Muller in 1954 ([29]) and the decoding method was due to Reed in 1954 [31]. Its decoding method is easy; hence, it has been used in several occasions, for instance, during 1969–1977, all of NASA's (USA) *Mariner* class deep-space probes were equipped with a Reed–Muller code (i.e., $RM(5,1)$, see following definition).

Recall from Section 5.1 that a linear coding theory has the following diagram

$$\text{message space } \mathbf{F}_q^k \xrightarrow{\sigma_1} \text{function space} \xrightarrow{\sigma_2} \text{word space } \mathbf{F}_q^n.$$

The first map σ_1 is injective. Thus, we use functions to rename the messages, and the second map σ_2 is an injective map with the image of $\sigma_1(\mathbf{F}_q^k) =$ the code space. Usually, the map σ_2 is an evaluation map which evaluate a function f at an ordered n-tuple of points $(P_1, \ldots, P_n)$. Thus, it maps a function f to $[f(P_1), \ldots, f(P_n)] \in \mathbf{F}_q^n$. Note that $\sigma_2\sigma_1$ will send the message space to the word space; certainly, we do not want to send any non-zero message to zero. Thus we require that the composition $\sigma_2\sigma_1$ is an injective map on the message space.

We shall use the following defined $P(m,r)$ as the function space. Let $P(m,r)$ denote the set of all polynomials of degree $\leq r \leq m$ in m variables $(x_1, \ldots, x_m)$ over $\mathbf{F}_2$. Note that computation wise, over $\mathbf{F}_2$, we always have

$$x^2 = x.$$

Hence, all monomials can be reduced in the computational sense to multiples of distinct x_i.

Let us consider $\mathbf{F_2}^m$ as the ground vector space. We have a simple way to represent $P(m,r)$. Let $n = 2^m$. Then there are $n = 2^m$ vectors (or points) in $\mathbf{F_2}^m$ and $(v_0, v_1, \ldots, v_{n-1})$ denote a list of all the 2^m binary vectors in $\mathbf{F_2}^n$ in some order. Then, for each $f \in P(m,r)$, we may define $f(v_j) = h(a_{j1}, \ldots, a_{jm})$, where $v_j = [a_{j1} \cdots a_{jm}]$. Furthermore, we may use integers n_j to represent v_j as we define $n_j = \sum 2^{(i-1)a_{ji}}$. We have the following example of using integers to represent point in F_2^n.

Example C2: Let us consider $m = 4$. Then, we have the following Table C.1 for the values of polynomials x_1, x_2, x_3, x_4. Numerically, x_i goes to 2^{i-1} and the coefficients a_1, a_2, a_3, a_4 in the table determine the polynomial $\sum (x_i)^{a_i}$ which in turn goes to $\sum 2^{(i-1)a_i}$, where $a_i = 1$ or 0.

Table C.1.

	0,	1,	2,	3,	4,	5,	6,	7,	8,	9,	10,	11,	12,	13,	14,	15
x_1	0,	1,	0,	1,	0,	1,	0,	1,	0,	1,	0,	1,	0,	1,	0,	1
x_2	0,	0,	1,	1,	0,	0,	1,	1,	0,	0,	1,	1,	0,	0,	1,	1
x_3	0,	0,	0,	0,	1,	1,	1,	1,	0,	0,	0,	0,	1,	1,	1,	1
x_4	0,	0,	0,	0,	0,	0,	0,	0,	1,	1,	1,	1,	1,	1,	1,	1

For the polynomials, we use the formula $(f + g)(v_j) = f(v_j) + g(v_j)$ to find their values. In this way, we may construct a table for the values of all polynomials. ∎

Furthermore, we get a binary vector of length $n = 2^m$ via the mapping $\sigma_2(f) = (f(v_0), \ldots, f(v_{n-1}))$. The set of all vectors obtained in this way is called the *rth order Reed–Muller code* of length $n = 2^m$, or $RM(m, r)$ for short. It is easy to see that $RM(m, r)$ is a linear code $[n, k, d]$ (in fact, it is a cyclic code). We find the parameters k, d.

Proposition C.2. *We have*

$$k = 1 + C_1^m + \cdots + C_r^m.$$

Proof. The number of monomials of degree j which are multiples of distinct x_i is clearly C_j^m; hence, the number of monomials of degree $\leq r$ is clearly

$$1 + C_1^m + \cdots + C_r^m.$$

Therefore, we need only to show that any non-zero polynomial f will not be sent to zero vector in F_2^n by σ_2, i.e., $\sigma_2(f) = (f(v_0), f(v_1), \ldots, f(v_{n-1})) \neq (0, 0, \ldots, 0)$. Suppose that $h(x_1, \ldots, x_m) = f$ is a linear combination of those monomials and $f(v_i) = 0$ for all i. If $h(0, x_2, \ldots, x_m) \neq 0$, then we can find values for $x_2, \ldots, x_m$ so that it is not zero by mathematical induction. If $h(0, x_2, \ldots, x_m) = 0$, then $h(x_1, \ldots, x_m) = x_1 g(x_2, \ldots, x_m)$ and $h(1, x_2, \ldots, x_m) = g(x_2, \ldots, x_m)$, so that we can find values for $x_2, \ldots, x_m$ such that $h(v_i)$ is not zero by mathematical induction. In any case, we have a contradiction if $\sigma_2(f) = (0, 0, \ldots, 0)$. □

Corollary C.3. *The vector space of all polynomials in m variables of degree less than or equal to r and having all terms with degree in each x_i being one or zero is isomorphic to $RM(m, r)$.*

Proposition C.4. *We have the minimal distance* $d = 2^{m-r}$.

Proof. We first show that $d \leq 2^{m-r}$. Let us use the notations of the proof of the preceding proposition. Let $h = x_1 x_2 \cdots x_r$. Then, $f = 0$ whenever $x_1 = 0$ or $x_2 = 0$ or $\cdots$ or $x_r = 0$. Using set-theoretic inclusions to compute the number of zeroes among $v_0, \ldots, v_{n-1}$, we may conclude that it is $2^m - 2^{m-r}$. There are at least $2^m - 2^{m-r}$ zeroes among $v_0, \ldots, v_{n-1}$. Therefore, the minimal weight, hence the minimal distance, is at most 2^{m-r}.

We wish to prove that $d \geq 2^{m-r}$. Let us consider a polynomial $h(x_1, x_2, \ldots, x_m) \neq 0$, so every variable appears at most once in every term. We have $h(0, x_2, \ldots, x_m) = 0 \implies h(x_1, \ldots, x_m) = x_1 g(x_2, \ldots, x_m)$. Similarly, $h(1, x_2, \ldots, x_m) = 0 \implies h(x_1, \ldots, x_m) = (x_1 - 1)p(x_2, \ldots, x_m)$. After we try every x_i, then we conclude that either there is an x_i, say x_1, such that (1) $h(0, x_2, \ldots, x_m) \neq 0$ and $h(1, x_2, \ldots, x_m) \neq 0$ or (2) $h = \prod_{i=1}^{m}(x_i - \delta_i)$, where $\delta_i = 0$ *or* 1. The second case happens only if $r = m$, and $2^{m-r} = 1$, our proposition is certainly true. In the first case, both $h(0, x_2, \ldots, x_m) = g(x_2, \ldots, x_m) \neq 0$ and $h(1, x_2, \ldots, x_m) = p(x_2, \ldots, x_m) \neq 0$. By induction on the number of variables, we conclude that $g(x_2, \ldots, x_m)$ and $h(x_2, \ldots, x_m)$ have at most $2^{m-1} - 2^{m-1-r}$ zeroes. Since the sets of zeroes for $g(x_2, \ldots, x_m)$ and $p(x_2, \ldots, x_m)$ are with $x_1 = 0$ or $x_1 = 1$, they are disjoint. Therefore, f has at most $(2^{m-1} - 2^{m-1-r}) + (2^{m-1} - 2^{m-1-r}) = 2^m - 2^{m-r}$ zeroes. So the minimal weight, hence the minimal distance, is at least 2^{m-r}. □

Proposition C.5. *We have* $RM(m, r)^{\perp} = RM(m, m - r - 1)$.

Proof. Let us first compute the dimensions:

$$\dim(RM(m, r)^{\perp}) = 2^m - (1 + C_1^m + \cdots + C_r^m)$$

$$= C_{r+1}^m + \cdots + C_{m-1}^m + 1$$

$$= C_{m-r-1}^m + \cdots + C_1^m + 1$$

$$= \dim(RM(m, m - r - 1)).$$

Now, we only need to show that every monomial in $RM(m, m - r - 1)$ is in $RM(m, r)^{\perp}$, i.e., with $f = (\prod_i x_i)(\prod_j x_j)$, where $(\prod_i x_i) \in RM(m, m - r - 1)$ and $(\prod_j x_j) \in RM(m, r)$; the following sum is zero always (note that after reducing the total degree of f by $x_i^2 = x_i$, we may take f to be a product of less than or equal to $m - 1$ distinct variables):

$$\sum_{i=0}^{n-1} (f(v_i)) = 2^{m-\deg(f)} = 0 \quad \text{mod} \quad 2.$$

The above equation is obvious. $\qquad\square$

The method of decoding RM codes is the *threshold decoding* described in Section C.4. The reader is referred to [31].

C.3. Constructing New Code from Old Codes

C.3.1. *Extend, Puncture, Shorten, Lengthen, Expurgate, and Augment a Code*

Let us take an example to explain the meanings of the operations. Let C_i be a linear code with the following check matrix H_i. Let H_0 be given as follows. Let C_0 be a cyclic code with generating polynomial $g_0(x) = x^3 + x + 1$.

$$H_0^T = \begin{bmatrix} 1 & 0 & 1 & 0 & 1 & 0 & 0 \\ 1 & 1 & 1 & 1 & 0 & 1 & 0 \\ 0 & 1 & 1 & 1 & 0 & 0 & 1 \end{bmatrix},$$

Extend: Any code can be extended by annexing additional check symbols. The resulting code is not cyclic in general. We have the following example with C_2 be given by the following check matrix"

$$H_2^T = \begin{bmatrix} 1 & 0 & 1 & 0 & 1 & 0 & 0 & 0 \\ 1 & 1 & 1 & 1 & 0 & 1 & 0 & 0 \\ 0 & 1 & 1 & 1 & 0 & 0 & 1 & 0 \\ 1 & 1 & 1 & 1 & 1 & 1 & 1 & 1 \end{bmatrix}.$$

Puncture: Any code can be punctured by deleting some of its check symbols. We have an example of $C_2 \to C_0$.

Expurgating: A cyclic code generated by the polynomial $g(x)$ can be expurgated to form another cyclic code by multiplying any additional factor into the generated polynomial. The most common expurgate code is the code C_1 generated by $g(x)(x-1)$. We have the following example $C_0 \to C_1$.

$$H_1^T = \begin{bmatrix} 1 & 0 & 1 & 0 & 1 & 0 & 0 \\ 1 & 1 & 1 & 1 & 0 & 1 & 0 \\ 0 & 1 & 1 & 1 & 0 & 0 & 1 \\ 1 & 1 & 1 & 1 & 1 & 1 & 1 \end{bmatrix}.$$

Augment: A cyclic code generated by the polynomial $g(x)$ may be augmented into another cyclic code of the same length whose generator polynomial is a factor of $g(x)$. The most common augmented code is the one generated by $g(x)/(x-1)$ (we assume that it is allowed, i.e., $g(1)=0$). We have the example $C_1 \to C_0$ with C_0 a cyclic code generated by $g_0(x) = g_1(x)/(x-1)$.

Lengthen: If the generator polynomial of an original binary cyclic code of length n contains the factor $(x+1)$, then this cyclic code can be lengthened to a linear code of length $n+1$ by annexing the vector $[11 \cdots 1]$ to the code's generator matrix. We have the example $C_1 \to C_2$.

Shorten: Any code can shortened by deleting message symbols. We have the example $C_2 \to C_1$ with C_1 a cyclic code.

To summarize the preceding discussion, we have the following figure, **Table C.2**.

C.3.2. *Direct Product of Cyclic Codes*

See Section 3.5.

C.3.3. *Concatenated Codes and Justensen Codes*

The concept of *concatenated codes* was introduced by Forney [22] in 1966. Let us consider a simple example. Suppose that we send out an e-mail which consists of several sentences in certain set of letters. We may first encode the letters in certain way. Say we have three letters, we may encode them in $\mathbf{F}_{2^5}$ as

$$00000$$
$$11100$$
$$00111.$$

Note that the minimal distance of the three letters is 3, and we expect to correct 1 error. Then, we may use some encoding scheme, say RS-code, to encode the whole e-mail. The way to encode the letters will be called the *inner code*, and the way to encode the e-mail will be called the *outer code*. In general, we may encode any message by a code C_1 (like the encoding of all letters) which is called the *inner code*, and then encode the code words C_1 by another code C_2 which is called the *outer code*. The most important development along this direction is the *Justensen codes* [28].

C.4. Threshold Decoding

We may use the *threshold decoding* to handle the problem of decoding several codes. We illustrate the threshold decoding by the following example.

Example C2: Let C a $[7,3]$ binary linear code with the check matrix H

$$H = \begin{bmatrix} 1 & 1 & 0 & 1 & 0 & 0 & 0 \\ 1 & 0 & 1 & 0 & 1 & 0 & 0 \\ 0 & 1 & 1 & 0 & 0 & 1 & 0 \\ 1 & 1 & 1 & 0 & 0 & 0 & 1 \end{bmatrix}.$$

Let $c = [c_0, c_1, \ldots, c_6]$ be the code word sent and $r = [r_0, r_1, \ldots, r_6] = c + e$ be the received word, where $e = [e_0, e_1, \ldots, e_6]$ is the error vector. This code may correct 1 error. We shall consider the code in a different way. Analytically, the check matrix produces the following system of equations satisfied by c_0:

$$c_0 = c_1 + c_3$$

$$c_0 = c_2 + c_4$$

$$c_0 = c_5 + c_6.$$

Now, let us assume the channel is *binary symmetric* with error probability $\wp < \frac{1}{2}$. Then, we have

$$c_0 = \begin{cases} r_1 + r_3 & : & \text{if } e_1 + e_3 = 0 \\ r_2 + r_4 & : & \text{if } e_2 + e_4 = 0 \\ r_5 + r_6 & : & \text{if } e_5 + e_6 = 0 \\ r_0 & : & \text{if } e_0 = 0. \end{cases}$$

If there is one error, then we may use *majority vote* to decide which value is correct for c_0. If two errors are allowed, and there is a 2-2 tie on the vote, there is one vote with $e_0 = 0$ since $e_0 = 0$ is with a probability $1 - \wp$ and $e_i + e_j = 0$ is with a probability $(1-\wp)^2 + \wp^2 = 1 - \wp - \wp(1 - 2\wp) < 1 - \wp$, so it is in favor of the vote with $e_0 = 0$, i.e., $c_0 = r_0$. ∎

Reed obtained a generalization of the above technique to RM codes (cf. [31]) which we illustrate by the following example. Later on, the method was applied to many codes and named *threshold decoding*. The interested reader is referred to MacWilliams and Sloane [7].

Example C3: Let us consider $RM(3,1)$ which is a $[8,4,4]$- code. We have the following Table C.2.

Table C.2.

	0,	1,	2,	3,	4,	5,	6,	7,
m_0	1,	1,	1,	1,	1,	1,	1,	1,
m_1	0,	1,	0,	1,	0,	1,	0,	1,
m_2	0,	0,	1,	1,	0,	0,	1,	1,
m_3	0,	0,	0,	0,	1,	1,	1,	1,

We have the following method of decoding.

Threshold Decoding: Let the original message be $[m_0, m_1, m_2, m_3]$, the code word be $c = [c_0, c_1, \ldots, c_7] = m_0[1, 1, \ldots, 1] + m_1[0, 1, 0, 1, 0, 1, 0, 1] + m_2[0, 0, 1, 1, 0, 0, 1, 1] + m_3[0, 0, 0, 0, 1, 1, 1, 1]$ and the received word be $r = [r_0, r_1, \ldots, r_7]$. Then, the relations between m_i, c_j are

$$c_0 = m_0, c_1 = m_0 + m_1, c_2 = m_0 + m_2, c_3 = m_0 + m_1 + m_2, c_4 = m_0 + m_3,$$

$$c_5 = m_0 + m_1 + m_3, c_6 = m_0 + m_2 + m_3, c_7 = m_0 + m_1 + m_2 + m_3.$$

The above relations can be written as follows:

$$m_1 = c_0 + c_1 = c_2 + c_3 = c_4 + c_5 = c_6 + c_7,$$

$$m_2 = c_0 + c_2 = c_1 + c_3 = c_4 + c_6 = c_5 + c_7,$$

$$m_3 = c_0 + c_4 = c_1 + c_5 = c_2 + c_6 = c_3 + c_7.$$

Clearly, we do not know c_i's and only know r_i's. Assume that there is at most 1 error. Let us consider the value of m_1, the values of m_2, m_3 can be considered similarly. Then, at least three of the four values of $r_0 + r_1, r_2 + r_3, r_4 + r_5, r_6 + r_7$ must be correct. Therefore, a majority vote will decide the correct value of m_1. If there are two errors, then the vote might be tied. In that case, the value is indeterminate. After we find the values for m_1, m_2, m_3, we are left to find m_0. Note that

$$[c_0, c_1, \ldots, c_7] = m_0[1, 1, \ldots, 1] + m_1[0, 1, \ldots, 0, 1] + m_2[0, 0, 1, \ldots, 1, 1]$$

$$+ m_3[0, 0, \ldots, 1, 1].$$

Once we find the correct values of m_1, m_2, m_3, let us compute

$$[r_0, r_1, \ldots, r_7] + m_1[0, 1, \ldots, 0, 1] + m_2[0, 0, 1, \ldots, 1, 1] + m_3[0, 0, \ldots, 1, 1].$$

The result of the above sum should be close to either $[1, 1, \ldots, 1]$ or $[0, 0, \ldots, 0]$. A majority vote should give us the correct value of m_0 (except the case of a 4-4 tie, then the value of m_0 is indeterminate). ∎

C.5. Soft-Decision Decoding

The practice of transmitting signals electronically involves wave functions through a medium. After a white noise is added to the wave function, there is a distortion to the signals. Usually, one cannot decide for certain if the signal s is 0 or 1 (we only consider binary bits). One can associate reliability probabilities p_0 and p_1 (note that $p_0 + p_1 = 1$) to the particular signal s being 0 and 1. We usually decide the bit is 0 if $p_0 \geq 1/2$ and 1 if $p_0 \leq 1/2$. We consider that if p_0 is close to 1, then it is reliable to assign the signal to 0, and if it is close to 0, then it is reliable to assign the signal to 1. If it is close to $1/2$, then the reliability is weak. One may use the values of p_0 and p_1 to decode the sequence of signals. This is what is called *soft-decision* decoding. On the other hand, if we only use the bits, without considering their reliability probabilities, to decode then it is called *hard-decision* decoding. So far in this book, we only consider hard-decision decoding.

Let us consider that the original message is $c = (c_0 c_1 \cdots c_n)$ and the received word is $r = (r_0 r_1 \cdots r_n)$ with the reliabilities sequence $(p_0^{(1)} \cdots p_0^{(n)})$. The Forney's *generalized minimum distance* (GMD) decoder [22] is essentially to erase the $d - 1$ least reliable symbols and then use an algebraic decoder to correct the remaining received word.

A more interesting method is to use the reliability probabilities $(p_0^{(1)} \cdots p_0^{(n)})$. Let us consider a 3 repetition code which for certain bit produces the reliability probabilities $(0.51, 0.51, 0.12)$ for $(0, 0, 0)$. A hard-decision (which we had considered so far) decoder will first make a decision that the bit received are $(0, 0, 1)$ (since $0.51 > 0.50, 0.12 < 0.50$) and then use the majority rule to decide the original bit is 0. A soft-decision decoder will take the reliability probabilities into consideration. It would take the average reliability probability as the deciding factor; the average reliability probability is

$$\frac{1}{3}(0.51 + 0.51 + 0.12) = 0.38.$$

Therefore, the original bit is more likely to be 1. In general, it can be shown that the soft-decision decoder is better.

C.6. Turbo Decoder

In 1993, at the IEEE International Conference on Communications in Geneva, a pair of French engineers, Claude Berrou and Alain Glavioux, announced their *turbo Codes* [17] which come very close to Shannon's theoretical results. It was surprising news to the world of coding theory.

Berrou likens the code to a crossword puzzle in which one would-be solver (decoder) receives the "across" clues and the other receives the "down" clues. After the decoders update their proposed solutions, they compare notes and update their solutions again. They continue this process until either they reached a limited number of times (say, 18 times) and stopped or they reach a consensus about the original message.

C.7. Low-Density Parity Check (LDPC) Codes

There is another code which rivals the turbo codes, the low-density parity check (LDPC) Codes [23] which was created by Robert Gallager, who was a Ph. D. student at MIT, in 1958. This code is sometimes called Gallager code. First, Gallager used *sparse matrices* for the generator matrices. Second, Gallager used a decoder for every bit and let the decoders talk among themselves and thus created a huge rumor mill with thousands or tens of thousands of talkers. The patent right of LDPC was held by Codex Corp. until it expired without ever being used. One of the reasons was it was technically infeasible to create the rumor mill in the 1950s.

Berlekamp's Decoding Algorithm

Berlekamp's decoding algorithm is the fastest decoding method for Reed–Solomon codes today. In industry, it is the main tool to decode the Reed–Solomon codes. Recall that in Section 3.2 we define the error-locator polynomial $\sigma(x)$ and the error-evaluator polynomial $\omega(x)$ as follows. Let $M =$ the set of error locations $\{\gamma^j\}$, then

$$\sigma(x) = \prod_{i \in M} (1 - \gamma^i x) = 1 + \cdots ,$$

$$\omega(x) = \sum_{i \in M} e_i \gamma^i x \prod_{j \in M \setminus i} (1 - \gamma^j x).$$

Let

$$r(x) = r_0 + r_1 x + \cdots + r_{n-1} x^{n-1} = \text{the received word},$$

$$S(x) = \sum_{i=1}^{2t} r(\gamma^i) x^i.$$

Then, we have the following equation:

$$S(x)\sigma(x) = \omega(x) \bmod x^{2t+1}.$$

We may slightly modify the above equation as

$$(1 + S(x))\sigma(x) = (\sigma(x) + \omega(x)) \bmod x^{2t+1}.$$

We may abuse the notation to write $\sigma(x) + \omega(x)$ as $\omega(x)$ (for the sake of induction later on) and get the following *key equation*:

$$(1 + S(x))\sigma(x) = \omega(x) \bmod x^{2t+1}.$$

A central problem in coding theory is given $S(x)$, how do we quickly produce $\sigma(x)$ and $\omega(x)$ with $deg(\sigma(x), deg(omega(x)) \leq t$? One of the simple methods is to look at the conditions and consider

$$S(x) = s_1 x + s_2 x^2 + \cdots + s_{2t} x^{2t},$$

$$\sigma(x) = 1 + a_1 x + a_2 x^2 + \cdots + a_t x^t,$$

where s_i are known and a_i are unknown. Since $deg(\omega(x)) \leq t$, we set the coefficients of $x^{t+1}, \ldots, x^{2t}$ in the product of $S(x)\sigma(x)$ to be zeroes and get the following linear equations:

$$s_t a_1 + s_{t-1} a_2 + \cdots + s_1 a_t + s_{t+1} = 0,$$

$$s_{t+1} a_1 + s_t a_2 + \cdots + s_2 a_t + s_{t+2} = 0,$$

$$\cdots\cdots\cdots$$

$$s_{2t-1} a_1 + s_{2t-2} a_2 + \cdots + s_t a_t + s_{2t} = 0.$$

Since we know that there are coefficients $\{a_i\}$ satisfy the above equations, the above system of equations is consistent and produces a solution for indeterminates $a_1, \ldots, a_t$. We just solve it, find $\sigma(x)$, and take $\omega(x) = S(x)\sigma(x) \bmod x^{2t+1}$. However, this method of solving systems of equations is slow. In this section we reproduce the original method of Berlekamp [4] which is still the fastest method today.

D.1. Berlekamp's Algorithm

Note that we already know the existence of $\sigma(x)$ and $\omega(x)$; it follows from Proposition 2.56 that they are unique. The only problem is how to find them fast. Note that we know $deg(\sigma(x)) \leq t$, $deg(\omega(x)) \leq t$ and $deg(S(x)) \leq 2t$. Berlekamp's idea is not to solve the above system of linear equations directly, rather to find a sequence of $\sigma^{(k)}(x)$ and $\omega^{(k)}(x)$ with $\sigma^{(2t)}(x) = \sigma(x)$ and $\omega^{(2t)}(x) = \omega(x)$. The *key equation* is generalized to a sequence of equations of the following form:

$$(1 + S(x))\sigma^{(k)}(x) = \omega^{(k)}(x) \bmod x^{k+1}, \qquad (D1_k)$$

with the degree restrictions polynomials $deg(\sigma^{(k)}(x)), deg(\omega^{(k)}(x)) \leq \frac{k+1}{2}$. We define $\sigma^0 = 1$, inductively; suppose that we have constructed $\sigma^{(k)}(x), \omega^{(k)}(x)$, we want to define $\sigma^{(k+1)}(x), \omega^{(k+1)}(x)$. Let us look at one more term as follows:

$$(1 + S(x))\sigma^{(k)}(x) = \omega^{(k)}(x) + \Delta^{(k)} x^{k+1} \bmod x^{k+2}, \qquad (E_k)$$

where $\Delta^{(k)}$ is the next coefficient and a number of the above equation. If $\Delta^{(k)} = 0$, then we may take $\sigma^{(k+1)}(x) = \sigma^{(k)}(x)$ and $\omega^{(k+1)}(x) = \omega^{(k)}(x)$ and continue our inductive process of construction. Suppose $\Delta^{(k)} \neq 0$. Then, it is complicated to define them. We have to do something more. We introduce two more functions $\tau^{(k)}$ and $\gamma^{(k)}$. Let us define them by the following equations:

$$\Delta(\sigma^{(k)}) = \sigma^{(k)} - \sigma^{(k+1)} = \Delta^{(k)} x \tau^{(k)},$$

$$\Delta(\omega^{(k)}) = \omega^{(k)} - \omega^{(k+1)} = \Delta^{(k)} x \gamma^{(k)}.$$

Inductively, after we define $\tau^{(k)}$ and $\gamma^{(k)}$, then we have to define $\tau^{(k+1)}$ and $\gamma^{(k+1)}$. let us take $(E_k) - (D1_{k+1})$ and simplify. Then, we deduce the following equation:

$$(1 + S(x))\tau^{(k)} = \gamma^{(k)} + x^k \bmod x^{k+1}. \tag{$D2_k$}$$

Inductively, we define $\tau^{(k+1)}$ and $\gamma^{(k+1)}$ as follows. We may use one of the following two ways, if $\Delta^{(k)} = 0$, let

$$\tau^{(k+1)} = x\tau^{(k)}, \text{ and } \gamma^{(k+1)} = x\gamma^{(k)}, \tag{D3}$$

or if $\Delta^{(k)} \neq 0$, let

$$\tau^{(k+1)} = \frac{\sigma^{(k)}}{\Delta^{(k)}}, \text{ and } \gamma^{(k+1)} = \frac{\omega^{(k)}}{\Delta^{(k)}}. \tag{D4}$$

The critical things are to control the degrees of $\sigma^{(k)}(x), \omega^{(k)}(x)$. We wish that they are less than or equal to $\frac{k+1}{2}$.

Initially, Berlekamp adds two more integer-valued functions $D(k), B(k)$ and defines $\sigma^{(0)} = 1, \omega^{(0)} = 1, \tau^{(0)} = 1, \gamma^{(0)} = 0$, and integer-valued functions $D(0) = 0, B(0) = 0$. Inductively, we have two cases:

$$\text{Case 1} = \begin{cases} \Delta^{(k)} = 0 \quad \text{or} \\[2mm] \Delta^{(k)} \neq 0, D(k) > \dfrac{k+1}{2} \quad \text{or} \\[2mm] \Delta^{(k)} \neq 0 \text{ and } D(k) = \dfrac{k+1}{2}, \text{ and } B(k) = 0, \end{cases}$$

$$\text{Case 2} = \begin{cases} \Delta^{(k)} \neq 0 \text{ and } D(k) < \dfrac{k+1}{2} \quad \text{or} \\[2mm] \Delta^{(k)} \neq 0, D(k) = \dfrac{k+1}{2}, \text{ and } B(k) = 1. \end{cases}$$

In case 1, we define $\tau^{(k+1)}, \gamma^{(k+1)}$ by equation (D3) and set

$$(D(k+1), B(k+1)) = (D(k), B(k)).$$

In case 2, we define $\tau^{(k+1)}, \gamma^{(k+1)}$ by equation (D4) and set

$$(D(k+1), B(k+1)) = (k+1 - D(k), 1 - B(k)).$$

It is easy to see that $B(k)$ only takes values 0 or 1. We have the following propositions.

Proposition D.1. *We always have the following:* (1) $deg(\sigma^{(k)}) \leq D(k)$ *with equality if* $B(k) = 1$. (2) $deg(\omega^{(k)}) \leq D(k) - B(k)$ *with equality if* $B(k) = 0$. (3) $deg(\tau^{(k)}) \leq k - D(k)$ *with equality if* $B(k) = 0$. (4) $deg(\gamma^{(k)}) \leq k - D(k) - (1 - B(k))$ *with equality if* $B(k) = 1$.

Proof. See [4].   □

Proposition D.2. *For each* k, *we have*

$$\omega^{(k)}\tau^{(k)} - \sigma^{(k)}\gamma^{(k)} = x^k.$$

Proof. See [4].   □

Proposition D.3. *If* $\sigma^{(k)}$ *and* $\omega^{(k)}$ *are any pair of polynomials which satisfy*

$$\sigma^{(k)}(0) = 1 \quad and \quad (1+S)\sigma^{(k)} = \omega^{(k)} \mod x^{k+1}.$$

Let $D = \max\{\deg \sigma^{(k)}, \deg \omega^{(k)}\}$. *Then, there exist polynomials* u *and* v *such that*

$$u(0) = 1, \quad v(0) = 0,$$
$$\deg u \leq D - D(k), \quad \deg v \leq D - [k - D(k)],$$
$$\sigma^{(k+1)} = u\sigma^{(k)} + v\tau^{(k)},$$
$$\omega^{(k+1)} = u\omega^{(k)} + v\gamma^{(k)}.$$

Proof. See [4].   □

Proposition D.4. *If* σ *and* ω *are relative prime and* $\sigma(0) = 1$ *and* $(1 + S)\sigma = \omega \mod x^{k+1}$, *then*

(1) *Either* $\deg \sigma \geq D(k) + 1 - B(k) \geq D(k)$, *or* $\deg \omega \geq D(k)$, *or both.*
(2) *If* $\deg(\sigma) \leq \frac{k+1}{2}$ *and* $\deg(\omega) \leq \frac{k}{2}$, *then* $\sigma = \sigma^{(k)}$ *and* $\omega = \omega^{(k)}$.

Proof. See [4].   □

We let $k = 2t$, then we inductively construct σ, ω and finish the Berlekamp's algorithm.

References

[1] Matthew: Matthew's Gospel, AD 80–90.

[2] Lao Tzu: Tao Te Ching. The oldest excavated portion dated to late 4th century BC. One of the English translation Fall River Press.

[3] Schrödinger, E. *What is Life. The Physical Aspect of Living Cell.* Cambridge: At The University Press. New York: The MacMillan Company, 1945.

[4] Berlekamp, E.R. *Algebraic Coding Theory.* New York: McGraw-Hill, 1968.

[5] Berlekamp, E.R. (ed). *Key Papers in The Development of Coding Theory.* New York: IEEE press, 1974.

[6] McEliece, R.J. *The Theory of Information and Coding.* Encyclopedia of Mathematics and Its Applications, Vol. 3, Reading, Mass: Addison-Wesley, 1977.

[7] MacWilliams, J. and Sloane, N. *The Theory of Error-Correcting and Coding.* Amsterdam: North Holland Publishing Co. 1977.

[8] Pretzel, O. *Codes and Algebraic Curves.* Clarendon Press, Oxford, 1998.

[9] van Lint, J.H. *Introduction to Coding Theory.* Springer, 1999.

[10] Chevalley, C. *Introduction to the Theory of Algebraic Functions of One Variable,* AMS, Providence, RI, 1951.

[11] Hartshorne, R. *Algebraic Geometry.* Springer-Verlag, 1977.

[12] Kurosh, A.G. *The Theory of Groups,* Chelsea, New York, NY, 1955.

[13] Mumford, D. *Introduction to Algebraic Geometry.* Springer-Verlag, 1964.

[14] Van der Waerden, B. *Modern Algebra.* Frederick Unger Publ. Co. 1950.

[15] Walker, R.J. *Algebraic Curves,* Princeton University, Princeton, Dover reprint (1962).

[16] Zariski-Samuel *Commutative Algebra,* Vol I & II D. Van Nostrand Co. 1960.

[17] Berrou, C. and Glavieux, A. Near optimum error correcting coding and decoding: Turbo codes. *IEEE Trans. Comm.*, **44**(10): 1261–1271. October 1969.

[18] Burton, H.O. and Weldon, E.J. Cyclic product codes. *IEEE Trans. Inf. Theory*, **IT-11**: 433–439, 1965.

[19] Elias, P. Coding for noisy channels. *IRE Conv. Record*, part 4, 37–46. Einhoven University of Technology, 1993

[20] Duursma, I.M. Decoding codes from curves and cyclic codes. Ph.D. dissetation, Einhoven University of Technology, 1993.

[21] Feng, G.L. and Rao, T.R.N. A simple approach for construction of algebraic-geometric codes from affine plane curves. *IEEE Trans. on Inf. Theory*, 40, 1003–1012, 1994.

[22] Forney, G.D.Jr. Generalized minimum distance decoding. *IEEE Trans. on Inf. Theory*, **12**(2): 125–131, April 1966.

[23] Gallager, R.G. Low-density parity-check codes. *IRE Trans. Inf. Theory.*, **IT-8**: 21–28. January 1962.

[24] Ghorpade, S. and Datta, M. Remarks on Tsfasman–Boguslavsky conjecture and higher weights of projective Reed–Muller codes. In *Arithmetic, Geometry, Cryptography and Coding Theory*, Providence, RI: AMS, 2017, pp. 157–169.

[25] Goppa,V.D. A new class of linear error-correcting codes. *Probl. Inf. Trans.*, **6**: 207–21, 1970.

[26] Hartmann,C.R.P. and Tzeng, K.K. Generalizations of BCH bound. *Inf. Control*, **20**: 489–498, 1972.

[27] Ihara, Y. Congruence relations and Shimura curves. *Proc. symp. Pure Math.*, **33**(2): 291-311, 1979.

[28] Justensen, J. A class of constructive asymptotically good algebraic codes. *IEEE Trans. Inf. Theory*, **18**: 652–656, 1972.

[29] Muller, D.E. *Metric Properties of Boolean Algebra and Their Application to Switching Circuits*. Report No. 46, Digital Computer Laboratory, Univ. of Illinois, April 1954.

[30] Peterson, W.W. Encoding and error-correction procedures for the Bose-Chaudhuri codes. *IRE Trans. Inf. Theory*, October 1960.

[31] Reed, I.S. A class of multiple-error-correcting codes and the decoding scheme. *J. IRE Trans. Inf. Theory*, September 1954.

[32] Reed, I.S. and Solomon, G. Polynomial codes over certain finite fields. *J. Soc. Ind. Appl. Math.*, June 1960.

[33] Shannon, C.E. A mathematical theory of communication. *Bell Syst. Tech. J.*, **27**: 379–423, 623–656, 1948.

[34] Skorobogatov, A.N. and Vlăduţ, S.C. On the decoding of algebraic-geometric codes. *IEEE Trans. Inf. Theory*, **36**: 1051–1060, 1990.

[35] Sugiyama, Y. Kasahara, M. Hirasawa, S., and Namekawa, T. A method for solving key equation for decoding Goppa codes. *Inf. Control*, **27**: 87–99, 1975.

[36] Tsfasman, M.A., Vlăduţ, S.C., and Zink, T. Modular curves, Shimura curves and Goppa codes better then the Varshamov-Gilbert bound. *Math. Nachr.*, **109**: 21–28. 1982.

[37] Xing, C. Nonlinear codes from algebraic curves improving the Tsfasman-Vlăduţ-Zink bound. *IEEE Trans. Inf. Theory*, **49**(7): 1652–1657, 2003.

[38] Hasse, H. Zur Theorie der abstrakten elliptischen funktionenkörper I,II & III. *Crelle's Journal*, **175**: 193–208, 1936.

[39] Serre, J.P. Sur le nombre des points rationnels d'une courbe algebraic sur un corps fini. *C.R. Acad. Sc. Paris*, **296**: 397–402, 1983.

[40] Weil, A. Number of solutions of equations over finite field. *Bull. Amer. Math. Soc.*, **55**: 497–508, 1949.

Index